Synthesis Lectures on Engineering, Science, and Technology

The focus of this series is general topics, and applications about, and for, engineers and scientists on a wide array of applications, methods and advances. Most titles cover subjects such as professional development, education, and study skills, as well as basic introductory undergraduate material and other topics appropriate for a broader and less technical audience.

Niels-Ebbe Dam

An Introduction to Electrical Circuit Theory

Volume 1: A Mathematical Approach to Fundamental Electrical Quantities and Resistive Circuits

Springer

Niels-Ebbe Dam
Klampenborg, Denmark

ISSN 2690-0300 ISSN 2690-0327 (electronic)
Synthesis Lectures on Engineering, Science, and Technology
ISBN 978-3-032-34345-1 ISBN 978-3-032-34346-8 (eBook)
https://doi.org/10.1007/978-3-032-34346-8

This Springer imprint is published by the registered company Springer Nature Switzerland AG
The registered company address is: Gewerbestrasse 11, 6330 Cham, Switzerland

Preface

Welcome to a book on electrical circuit theory, which the author uses in teaching three courses at the Technical University of Denmark, DTU.[1] The content of the three courses is resistive circuits, DC circuits, and AC circuits, respectively. This corresponds to the three volumes of the book which are mentioned below. In addition, the author teaches two electrophysics courses that deal with electrostatics and magnetostatics, respectively. These five courses are synchronised and offered in the first year of the power engineering programme at DTU.

Volume 1: A Mathematical Approach to Fundamental Electrical Quantities and Resistive Circuits has 4 chapters.

Chapter 1 provides an overview of the first two volumes. It also provides some recommendations on how a new student can organise their education and how to work with the first two circuit theory courses corresponding to Volumes 1 and 2.

Chapter 2 reviews very fundamental mathematical topics as well as solving systems of equations using matrices and determinants.

Chapter 3 introduces fundamental electrical quantities such as current, voltage, electrochemical potential, resistance, and power with an emphasis on electrophysical understanding. This is followed by rules for calculating the sign of current, voltage, and power. Finally, voltage and current generators are introduced.

Chapter 4 is a comprehensive review of many different methods for analysing DC circuits with resistors and voltage generators.

Volume 2: DC Circuit Theory deals with capacitors, coils, diodes, transistors, and operational amplifiers. It discusses, among other things, the charging and discharging processes for capacitors and coils using first-order differential equations, simple circuits with diodes and transistors, and second-order circuits where there are both coils and capacitors. In addition, many different circuits with operational amplifiers are described.

[1] Technical University of Denmark, the Ballerup Campus, the Department of Engineering Technology and Didactics.

Volume 3: AC Circuit Theory deals with the mathematical description of AC current, voltage, and power. The chapter continues with the mathematical description of three-phase circuits. Finally, a detailed description of the single-phase transformer and its equivalent diagrams is provided. The last chapter deals with filter circuits.

Volumes 2 and 3 are planned for later publication.

A textbook for an electronics engineering programme would probably teach students to analyse complicated circuits with a lot of circuit elements using computer programmes. This book does not. It deals with circuits with relatively few circuit elements. However, it goes into depth with mathematical derivations and proofs of the basic relationships and analysis methods. At the same time, emphasis is placed on physical understanding down to the smallest detail. This is reflected in the tests that accompany this book. As a student, one learns to calculate everything manually using mathematics and at the same time understand the physics, which is necessary for power engineering education, where in many cases one considers circuits with relatively few circuit elements.

Some other textbooks come in one volume with over 1000 pages with so many examples that the actual theory is interrupted so often by examples that it can be difficult to follow the main idea. Furthermore, the very thick book has countless problems that the student cannot even manage to calculate. This may discourage new students at the beginning of their studies. This textbook differs from many other textbooks in several ways

- The book begins with a chapter with suggestions for new students, how to organise their education, and how to work with the circuit theory courses.
- The book has a chapter where one reviews basic mathematics and is introduced to slightly more advanced mathematics that will be used in circuit theory courses.
- The book has fewer pages than many other university-level textbooks. This has been achieved by reducing the number of examples and exercises to the minimum necessary to gain an understanding of how theory is used to solve problems. However, the mathematical and scientific level has not been reduced, quite the opposite. The book is divided into three volumes so that it is more manageable to the student. In this way, the author hopes that it is easier to get an overview of the material and easier to read the theory without being interrupted by a lot of examples and exercises. Only theory that is important for education and curriculum has been included. The book contains tests that are suitable for achieving a good scientific, mathematical, and physical understanding of the theory. It is recommended to calculate a few problems to see how theory is used to solve practical problems. The use of computer programmes can be introduced in connection with some of the experimental exercises.
- With the emergence of AI, this book is an effort to ensure that basic fundamental knowledge is provided to future engineers.

The Creation of the Book

The teaching material itself has been created over the many years that I have taught these courses, long before AI was invented.[2] I have produced all the figures myself (with the exception of a few photos). It is not that I have developed original theories. The theory reviewed can be found in many other textbooks.[3] The only original thing about the book is the figures and the way the theory is explained. In addition to this, there are, of course, also the points mentioned above where the book differs from many other books.

The book is written in accordance with the IUPAC standard for letter formatting.[4]

Klampenborg, Denmark

Niels-Ebbe Dam

---

[2] In preparation for printing this book, AI was only used to make linguistic corrections, In each case, the author verified that the corrections have not affected academic content.

[3] The author has drawn his inspiration from several different textbooks, for example:
James W. Nilsson, Susan A. Riedel: Electric Circuits.
Allan R. Hambley: Electrical Engineering.
E.V. Sørensen: Elementær Kredsløbsteori (Danish book in three volumes).

[4] The main text is set in Roman style, and numbers, chemical symbols, and units are also set in Roman style. *Quantities and the charge quantum are set in Italics.* Comma (,) is used as a decimal separator, and period is used as a full stop (.) and not as a thousand separator.

Acknowledgements

Regarding names of DTU institutes and departments: Please see the footnote.[5]

Mehdi Savaghebi is Professor and Head of the section "Energy Technology and Computer Science".[6] It was his idea that I should publish my teaching material in circuit theory courses as a book.

Boye C. Knutz is Associate Professor Emeritus.[7] He has been an invaluable help in the creation of this book. He has proofread Chaps. 3 and 4 and made valuable suggestions for these chapters.

Iver Mølgaard Ottosen is Associate Professor.[8] He is Mathematician and Physicist. He has been of great help in the publication of this book. He has proofread Chap. 2, which deals with mathematics, and made valuable suggestions for this chapter.

Thomas Hjort Jensen is currently Senior Executive Communications Officer (See Footnote 8). Previously he worked as Editor in the creation of the Danish National Encyclopaedia. He has been of great help with good advice on the technical aspects of the publication process and the communication with Springer Nature.

Alexander Smith graduated originally with a B.Mus. (Hons) and M.Mus. (Orchestra Conducting). He has spent a large part of his life in Australia and the USA and has proofread the English language in that capacity. He is currently completing his education as Power Engineer.

Associate Professor *Ashraf Fathi Khalil Sulayman*, Assistant Professor *Sam Roozbehani*, and Associate Professor Emeritus *Esben Larsen*. They have been very helpful (see Footnote 7).

---

[5] DTU = Technical University of Denmark

"Department of Engineering Technology and Didactics" is a DTU institute.

"Energy Technology and Computer Science" is a section of "Department of Engineering Technology and Didactics".

[6] This section is a part of the DTU institute "Department of Engineering Technology and Didactics".

[7] In the section "Energy Technology and Computer Science", DTU.

[8] In the "Department of Engineering Technology and Didactics", DTU.

Claus Schelde Jacobsen is Professor at DTU. He has been very helpful with regard to electrophysical questions.

Rasmus Post is Associate Professor at Aarhus University. He helped me get started using LaTeX.

Other people who have shown positive interest and support for the creation of the book:

Malene Kirstine Holst is Head of the institute "Department of Engineering Technology and Didactics", DTU.

Lars D. Christoffersen was previously Head of the same institute (see Footnote 8). Now, he is Dean of Undergraduate Studies and Student Affairs at DTU.

Axel Grøndahl Kristiansen, who was previously Deputy Director of the institute (see Footnote 8).

Christen Monberg and Anders Troi, both former Heads of our section (see Footnote 7).

The many students who have appreciated the teaching material and encouraged me to publish it in book form.

I would like to thank Editorial Director *Charles Glaser* from the Springer Nature book publishing company for his welcoming, positive, and encouraging interest in realising the project of this book. Also thanks to the production team led by Coordinator *Srinivasan Sankar*.

Most of all thanks to my lovely wife *Vibeke* and daughter *Maria* for their positive interest and patience!

Contents

Introduction

1

1.1 Overview of the Chapters

1.1.1 Chapter 1. Introduction

The first part of this chapter is a brief overview of the first ten chapters corresponding to volumes 1 and 2. These two volumes correspond to the first two courses in the first half year of the power engineer education. The second part of this chapter contains recommendations for how to work with the material (Fig. 1.1).

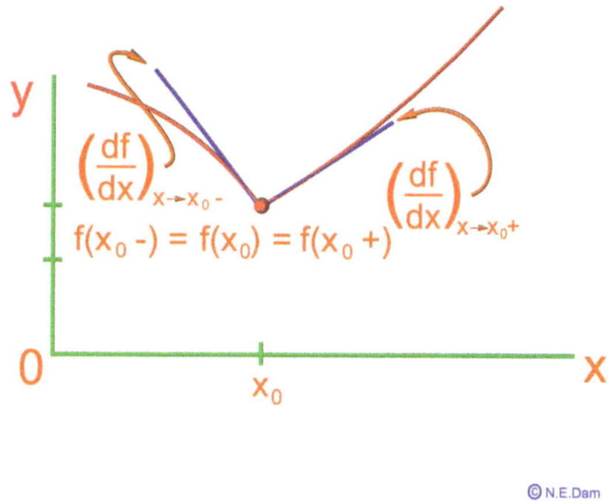

Fig. 1.1 Half-tangents for a continuous function at point x_o

N. Dam, *An Introduction to Electrical Circuit Theory*, Synthesis Lectures on Engineering, Science, and Technology, https://doi.org/10.1007/978-3-032-34346-8_1

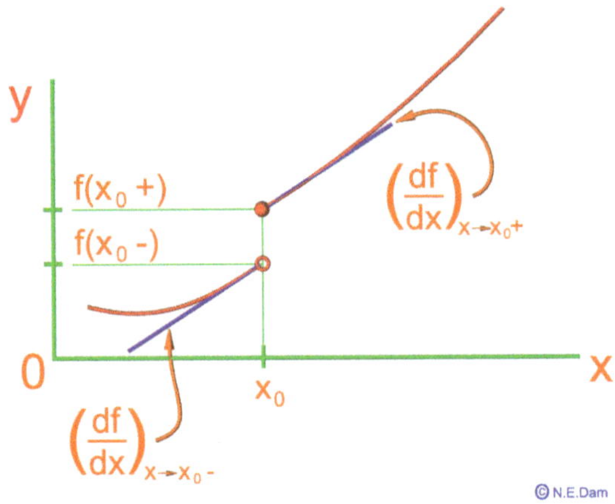

Fig. 1.2 Half-tangents for a discontinuous function at point x_0

1.1.2 Chapter 2. Mathematical Foundation

This chapter is a repetition of some of the most important mathematical concepts used in electrical circuit theory. It is very simple maths, only to ensure that everyone in the class has the necessary prerequisites to understand the chapters in this book. Some of the topics are:

- Brackets and fractions
- Exponential and power functions
- Differentiation and tangents (see, e.g., Fig. 1.2)
- Integration
- Two equations in two unknowns
- Three equations in three unknowns
- Differential equations of first and second order

When describing two and three equations with two and three unknowns, it is shown how the equations are put into matrix form and solved using determinants. As an example, a real electrical circuit is used, where we introduce the very important node voltage and mesh current methods.

1.1.3 Chapter 3. Current, Voltage, and Resistance

The chapter opens with a definition of electric current, electric voltage, and electric resistance, which are mutually related to each other through the current-voltage relationship, known as Ohm's law, if the resistance is constant, i.e. independent of current at constant temperature, see Fig. 1.3.

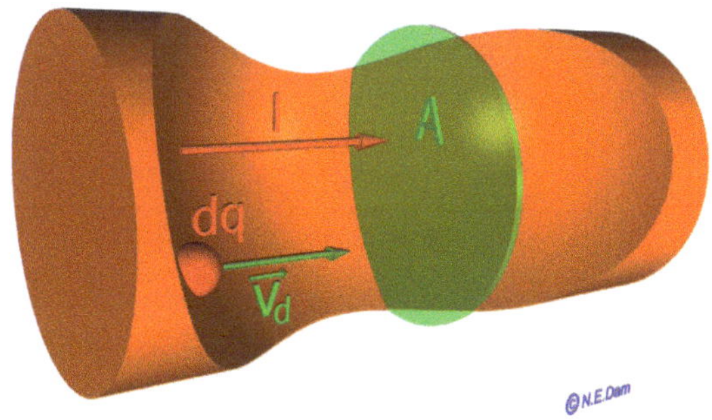

Fig. 1.3 Definition of electrical current. An amount of charge dq passes through a cross section area A of a wire

Next, a section explains how to determine the sign of the current through a circuit element and the sign of the voltage drop across a circuit element. This is called a configuration of a circuit element.

This is followed by a preliminary and short introduction to Kirchhoff's current law and Kirchhoff's voltage law. These two laws (and Ohm's law) form the basis for all the methods used to calculate currents and voltages in electrical circuits.

The chapter ends with two sections that describe definitions and equivalent circuits for voltage sources and current sources, respectively.

1.1.4 Chapter 4. Resistive Circuits

This chapter is the largest and, in a way, also the most important chapter in this book. The chapter describes various methods for calculating currents and voltages in electrical circuits with resistors. In addition, methods are described to simplify electrical circuits so that they become simpler to calculate and easier to construct.

The reason why the subject of resistive circuits is covered so extensively in this chapter is that all calculation methods and simplifications can also be applied to circuits that contain resistors, coils, and capacitors.

The chapter begins by describing the simplest circuits, namely the calculation of series and parallel connections of resistors, as well as voltage divider circuits and current divider circuits.

Then the Thévenin equivalent circuit is described. This circuit is used to simplify an electrical circuit to a series connection of a voltage source and a resistor, which is an advantage if one is only interested in what comes out of the electrical circuit. The equivalent circuit also exists in a version in which one simplifies an electrical circuit to a parallel connexion of a current source and a resistor.

This is followed by a section on the node voltage method, which is one of the most used methods to calculate electrical circuits. In this method, Kirchhoff's current law is combined with Ohm's law for each relevant node in the electrical circuit. This results in a number of

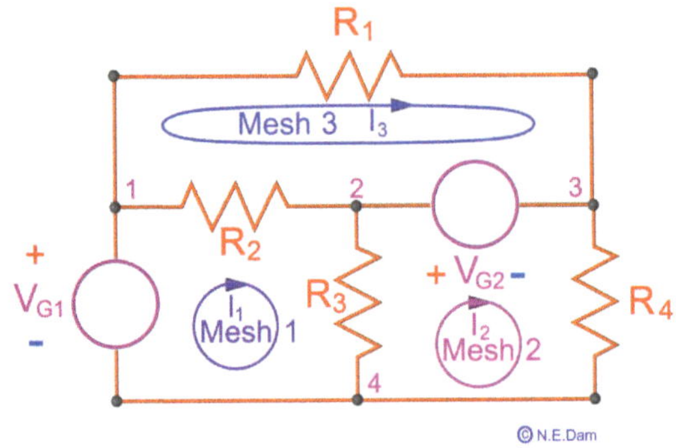

Fig. 1.4 Solving a circuit using the mesh current method

linear equations corresponding to the number of nodes in the circuit where the voltage, or more accurately the electric potential, is not known in advance. Linear equations are written in matrix form, after which the solutions are determined by calculating determinants for the system of equations.

Another widely used method for calculating electrical circuits is the mesh-current method. Here, Kirchhoff's voltage law is combined with Ohm's law for all relevant meshes in the circuit. This also results in a number of linear equations that are rewritten in matrix form and solved using determinants, as described for the node-voltage method. The result is mesh currents, which can be used to determine the potentials at the nodes of the circuit, see Fig. 1.4.

Sometimes, one comes across an electrical circuit, where three resistors form a triangle. This triangle is also called a Delta-connected circuit after the capital Greek letter Δ, which looks like a triangle. It will be shown how this triangle- or Delta-connected circuit can be transformed into a Y-connected circuit, i.e. a circuit of three resistors arranged in a large Y. This can lead to a simplification of the entire circuit, so that it is easier to calculate. The reverse transformation from a Y-connected circuit to a Delta-connected circuit is also shown. The chapter ends with a section dealing with different methods of measuring resistance with greater or lesser precision.

1.1.5 Chapter 5. Capacitive Circuits

This chapter starts with a definition of the concept of capacitance, which is a measure of how much charge the capacitor can contain when the voltage (or potential difference) between the two plates is 1 V; see Fig. 1.5. Subsequently, we consider the calculation of the capacitance for series or parallel connections of more capacitors. We also consider the influence on capacitance if an insulating material is inserted in between the two capacitor plates.

After this introduction, we derive the current-voltage relationships for the capacitor. For a resistor, we have the well-known relationship between the current through the resistor and voltage drop across the resistor, known as Ohm's law. Similarly, we also have a relationship

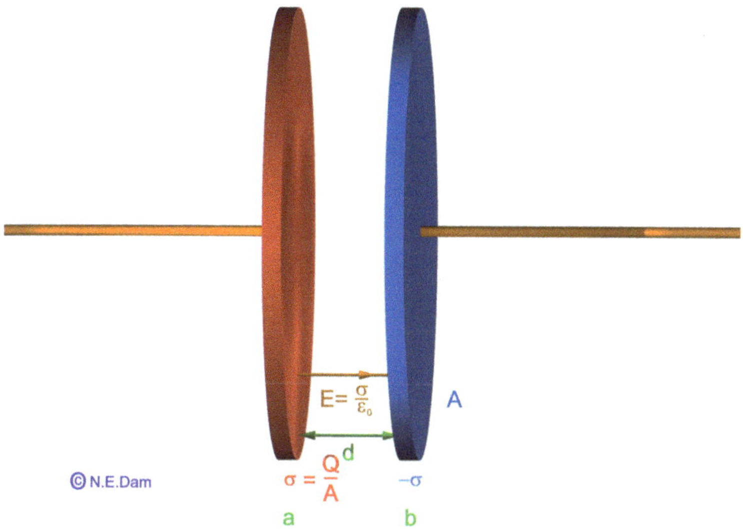

Fig. 1.5 A plate capacitor. A is the cross section or plate area, E is the electric field, Q is the charge on the positively charged plate, σ is the area charge density of charge and ϵ_o is the vacuum permittivity

between the current flowing into the capacitor and the potential difference between the two capacitor plates.

This is followed by a derivation of the current as a function of time for a capacitor that is charged and discharged by direct connection to a voltage source, where the voltage is a known function of time.

The chapter concludes with some sections that describe the charging and discharging of a capacitor connected to a voltage source through a resistor. In this connection, the voltage across the capacitor is found as a function of time. This voltage appears as the solution to a first-order homogeneous differential equation.

1.1.6 Chapter 6. Inductive Circuits

This chapter has a relatively long introduction, where we introduce some basic concepts within magnetism in order to better understand how a coil works. This is because elementary electrical engineering often appears earlier in an educational course than in a more detailed and comprehensive treatment of magnetism in a later electrophysics course. In the introduction to this chapter, we first look at sources of magnetic fields, then the magnetic flux, and Faraday's law of induction. On the basis of this, the operation of the coil and the basic current-voltage relationships for a coil are described (Fig. 1.6).

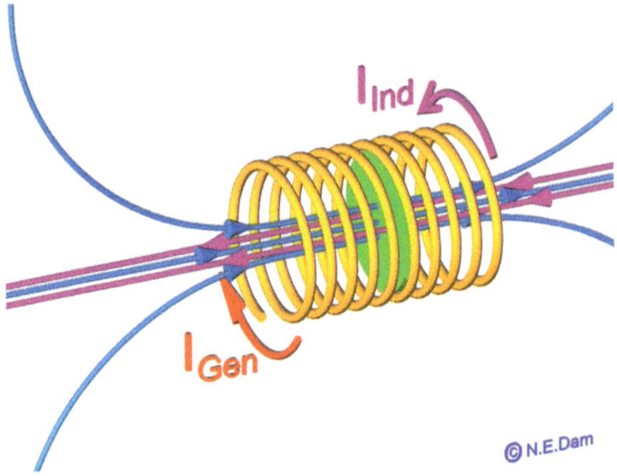

Fig. 1.6 A coil is passed through by a current I_{Gen}, where Gen stands for generator or source. The wiring between the coil and the current source is not shown. This current in the coil produces a magnetic field, which is symbolized by the blue magnetic field lines. Due to magnetic induction, the coil generates a current, an induced current, I_{Ind}, which is oppositely directed to the current from the current source. The induced current produces a magnetic field, symbolized by the purple field lines, which tries to counteract the magnetic field from I_{Gen}

This is followed by a derivation of the voltage as a function of time for a coil that is charged and discharged by current, when the coil is in direct connection to a current source, where the current is a known function of time.

The chapter concludes with some sections that describe the charging and discharging of a coil connected to a constant voltage source through a resistor. In this connection, the current through the coil is found as a function of time. This current appears as the solution to a first-order homogeneous differential equation.

1.1.7 Chapter 7. Second-Order Circuits

A second-order circuit contains both a capacitor and a coil, and typically also a resistor. The circuit elements can be connected in parallel or in series connection as shown in Fig. 1.7. The name Second-order Circuit is due to the fact that if one wants to determine the voltage across the capacitor, then this voltage is determined as the solution to a second-order differential equation.

A series or parallel connection between a capacitor and a coil is often called an oscillating circuit because the accumulated energy is constantly changing between the capacitor and the coil under the right conditions. At one instant there is a maximum voltage across the capacitor, which thus contains the accumulated energy as electrical potential energy. In the next moment, there is a maximum current through the coil, which thus contains the accumulated energy as magnetic potential energy.

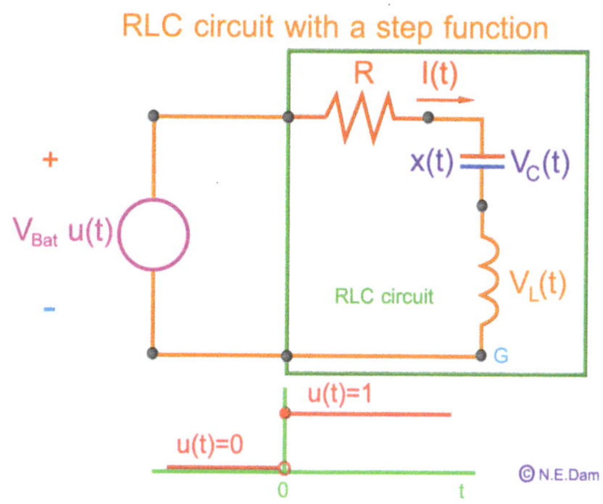

Fig. 1.7 A second-order circuit with a series connection of a resistor, a capacitor, a coil and a voltage source, which is turned on at time t = 0

Such circuits are of great practical importance, for example as filters that allow certain frequencies to pass through the filter, while other frequencies - such as noise - do not pass through the filter. Older radio receivers have thus used such a circuit, where the frequency corresponding to the radio station to which you want to listen passes through the oscillation circuit, while all other frequencies do not pass through.

The chapter shows how to set up the associated circuit equation, which, as mentioned, turns out to be a second-order differential equation. It shows how to solve this differential equation and how to plot the solution, which here is the voltage across the capacitor versus time for the circuit example shown in Fig. 1.7.

1.1.8 Chapter 8. Diode Circuits

A diode is a circuit element that in principle only allows current to pass one way, as shown in Fig. 1.8. However, certain types of diode allow the current, under certain circumstances, to be in the opposite direction, called the reverse direction.

This chapter deals only with the circuit theory associated with some important diode types. Semiconductor physics is treated in another chapter in electrophysics.

The first section covers the basic type of diode and its IV characteristic, which shows the current through the diode versus the applied voltage across the diode as shown in Fig. 1.8.

The Zener diode is mentioned, which allows current in the reverse direction if the applied voltage is sufficiently large in the reverse direction. Such Zener diodes are used in voltage supplies, where one can set a voltage that remains constant.

The next section deals with the photo-sensitive diode, which can be used in two ways:

1. as a diode that can measure brightness.
2. as a diode that produces voltage and current when illuminated, i.e. a solar cell.

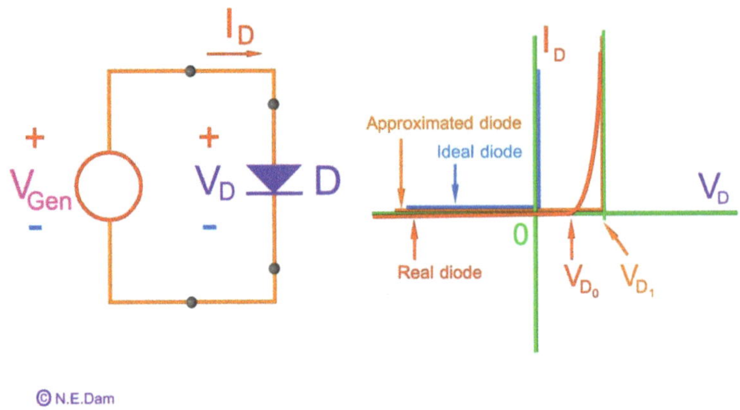

Fig. 1.8 A diode and its IV characteristic

For both applications, we see how to calculate the associated circuit using the curves in the IV characteristic.

This is followed by a small section with diodes that emit light, the Light Emitting Diodes, LEDs. They have an IV characteristic similar to the regular diode shown in Fig. 1.8, but with different values for the voltages V_{D0} and V_{D1}. LEDs are used as the well-known light-emitting diodes as well as in TV and computer screens.

At the end of the chapter, it is shown how diodes can be used to rectify alternating voltage into a pulsating direct voltage that can be "smoothed out" using a capacitor. Small rectifiers are used in small adapters for electronic devices and large rectifiers are used in the electricity supply network.

1.1.9 Chapter 9. Transistor Circuits

Roughly speaking, a transistor is a further development of a diode in the way a transistor has a third leg which is used to switch the current between the other two legs on and off. In Fig. 1.9, a current I_C flows through the transistor from terminal C (the Collector) to terminal E (the Emitter). This current can be turned on and off using a much smaller current flowing into terminal B (the Base). The transistor is used in not only digital circuits such as memory circuits, microprocessors, computers, smartphones, television, etc., but also in many analogue circuits and amplifier circuits.

The first section presents the circuit symbol and defines positive directions for the currents entering the transistor.

The next section shows the input characteristic, i.e., the IV characteristic of the diode from terminals B to E showing I_B versus the voltage drop V_{BE}.

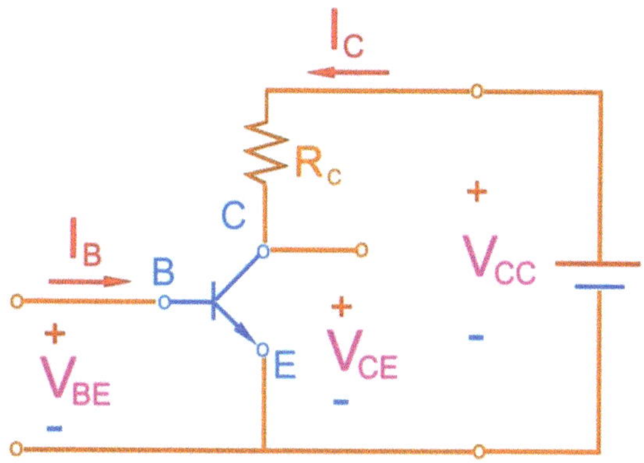

Fig. 1.9 Transistor circuit with a collector resistance. N.E.Dam ©

The third section describes the output characteristic showing the collector current I_C versus the voltage drop V_{CE} across the transistor from terminal C to terminal E. This IV characteristic is shown for different values of the base current I_B. The current gain, which is defined as the ratio $\beta = \frac{I_C}{I_B}$, can be read from the output characteristic.

The fourth section describes the three possible states of the transistor (cut-off, saturated, or active state). These three states correspond to three areas in the output characteristic.

The fifth section describes how to perform calculations on the circuit in Fig. 1.9. If you know V_{CC} and R_C, you can draw a load line in the output characteristic and thereby find the operating point which for a given value of I_B determines the value of I_C and V_{CE}.

The last two sections describe how to calculate curves that show I_C versus I_B and V_{CE} versus I_B, respectively.

1.1.10 Chapter 10. Operational Amplifier Circuits

An operational amplifier is an integrated voltage amplifier circuit that can amplify relatively small voltages up to a usable level. It can, for example, be a very small voltage (a few mV) from a thermometer or from a bio-potential electrode, which must be amplified by a factor of 1000 to obtain a signal that can be sent to a measuring instrument or to a microprocessor. By connecting resistors to the input and output terminals, the amplifier can also add and subtract voltages from different inputs, and much more. Figure 1.10 shows an example of an operational amplifier circuit.

After some introductory sections with definitions, a number of sections follow where analyses of various operational amplifier circuits are carried out. In these circuits, the operational amplifier is included in various resistance circuits. The combination of the resistors and the operational amplifier gives these circuits different characteristics.

A few examples:

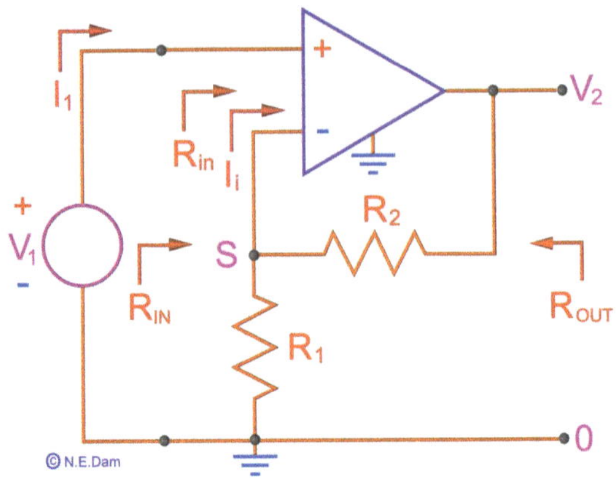

Fig. 1.10 The non-inverting amplifier circuit. The operational amplifier itself is the triangular diagram symbol. The terminals for voltage supply are not shown

Consider the non-inverting operational amplifier circuit in Fig. 1.10. In the corresponding section, it will be shown that one can calculate the output voltage to be $V_2 = (1 + \frac{R_2}{R_1})V_1$, where V_1 is the input voltage.

For the summing operational amplifier circuit, the output voltage is equal to the negative sum of the input voltages from 2, 3 or more inputs. The input voltages can be given individual gain factors determined by resistances in the circuit.

For the differential amplifier circuit, it will be shown that the output voltage becomes the voltage difference between two input times a gain factor determined by two resistors.

The chapter concludes with a comprehensive analysis of the instrumentation amplifier, a circuit with resistors, and three operational amplifiers.

1.2 How to Organize the Education

1.2.1 How to Get an Overview of the Education

This book is typically used in the first semester of electrical engineering education. Many of you reading this book are at the beginning of your university education. It is therefore very important to have a system in place that allows you to plan how to do the necessary work that will ensure success.

First, you have to make sure that you really want the subjects included in the education. You can talk to older students. You can look at concise descriptions of the subjects and see what subjects can be used in practice. When working your way through a theoretical course, it can be helpful to keep an overview of the subjects included in a course and the associated practical applications at all times. It can help maintain motivation.

You can get an overview of the entire education by writing the education's courses on each of its own small cards and placing them on a table (like a card solitaire). For each card, you can write:

- the title of the course;
- in which semester of the educational programme the course is normally taken in
- which day of the week is taught
- the most important topics
- practical applications

If, during education, you have to postpone a course, you can walk through these cards and get an overview of a revised study plan. You can also consider which optional courses should be included in the education. In this way, you can specialise in what you are most interested in and what you want as a graduation project.

The following two sections deal with two parameters that are important for a successful start-up and completion of the education:

- How to work appropriately in class or during lectures
- How to calculate and plan how much time to spend on each course in education

After these two sections, it is finally described how the courses are organised in the first semester of the electrical energy engineering programme at the DTU—Ballerup Campus.

1.2.2 How to Work Appropriately in Class or During Lectures

Of course, it is up to each individual student how they want to work with the education. When you work at home, you must try to arrange a workspace where you can work undisturbed for at least part of the day. Sometimes, it may be necessary to consider others. This also applies in a class or at a lecture, where others can disturb you or come to disturb others. The following is how you can unwittingly disturb others around you or the lecturer.

Disturbing others by talking to the person next to you
If you—while the lecture is in progress—talk or whisper to the person next to you, it can disturb the other students sitting around you. Thus, the other students cannot hear the lecturer and cannot concentrate on the subject. It can also disturb the lecturer if some of those being lectured to are inattentive and others have difficulty hearing the lecturer.

When you are working on solving problems, you are supposed to talk to the person next to you or the others around you. But when the lecturer indicates that she/he wants to continue the lecture after a break, you can help the lecturer by quickly stopping the conversation with the others around you.

Disturbing others by using your laptop in an inappropriate way.
There are three ways to use your laptop:

A. Keep the laptop closed while lecturing. Maybe you take notes on paper or just listen to the lecture. Some find it easier to take notes by hand as you can easily switch between writing text and drawing figures.
B. Use the laptop during lectures, but only to take notes. Here, it can be advantageous to have a small smart board where you can write and draw with a pen on a screen.
C. The laptop is used to surf the Internet and play games.

In all three cases, it is understood that the laptop is used when necessary to calculate the problems.

If all the students are A-students, we might get the best lecture and the most efficient teaching. In principle, there is no problem if there are some B-students. However, in practice some of the B-students swap between B- and C-behaviour, or they inspire other students to become C-students.

Some teachers don't care if some of the students are using their laptop to take notes or to update Facebook, etc. In general, it is not recommended to use the laptop during lectures. Other teachers do not allow open laptops during the lecture for many reasons, which will be discussed in the following.

(a) A-students looking towards the teachers whiteboard and the teacher might be disturbed if they see a lot of C-students with laptops running many different applications. It may cause some of the A- or B-students to leave the classroom simply because they cannot hear the teacher or they cannot concentrate on the lecture!
(b) The teacher becomes demotivated because the C-students do not listen attentively. Students may expect the teacher to be 100% focused and to give good lectures. It is also true the other way around. The teacher may expect that the students are 100% concentrated on the lecture. It might be difficult for the teacher to deliver the best possible lecture if the students pay more attention to their own laptop than to the lecture.
(c) The C-students send out a kind of signal indicating less motivation, less respect for the study, and less respect for the lecture. From experience, this signal may cause other students not to choose C-students when groups are formed. Hence, C-students may isolate themselves in groups that work poorly and can have difficulty understanding the subject.
(d) The C-students have a switch-mode behaviour between the lecture and their own laptop. Hence, they miss many details during the lecture and have a poorer understanding. The more active A-students may be irritated when the C-students ask questions because of their poorer concentration during the lecture. When the teacher spends time repeating theory for the C-students, it reduces the time that can be used for the A-students or used to explain more complicated parts of the curriculum in a better way. The teacher now faces two options, neither of them being acceptable:

i. The teacher keeps the pace at a level suited for the A-students. The A-students are now satisfied. The teacher gets through all the planned material before the end of the lecture, and everything is explained in full detail. However, there is now a group of C-students with a poor understanding, and this requires a lot of extra work for the teachers and for the assistant teachers after the lesson.
ii. The teacher lowers the pace of the lesson and repeats many times to ensure that the C-students also understand the subjects from the lecture. Now, it is difficult for the teacher to get to all the material planned for this day before the end of the lecture. The A-students complain about the slow progress and the many repetitions in the lecture. They may also complain about the very high pace just before the end of the lecture if the teacher tries to reach all the planned material.

1.2.3 The Workload

It is extremely important to set aside the necessary time each week.

How to calculate the workload
By an international definition, 60 ECTS points are equivalent to one year of "normal" work load outside the university. For example, it could be about 1620 h. Thus, 10 ECTS points becomes 1620/6 = 270 h and 5 ECTS becomes 135 h. If the university had the same number of holidays, etc. as outside the university, the weekly load could be e.g. 37 h. However, because of the very long summer break, the weekly load becomes 45 h if the university has 36 active weeks per year. Hence, it is not recommended to do additional work in addition to the study during these 36 weeks.

An example: A 5 ECTS course is completed in 13 weeks plus a couple of days in the following exam period. Then you have to spend 135 h in 13–14 weeks, or about 9–10 hours per week, when you are mentally fresh and not tired. This includes all work on this course, i.e. lectures, tests, assignments, and time to understand the material. This book is used (together with tests, assignments, and experimental exercises) at DTU for two 5 ECTS courses in the first semester. Therefore, 20 h per week must be allocated for these two courses!

How extra work may be combined with the study
As mentioned above, it is strictly not recommended to have extra work. If it is absolutely necessary for survival, then it is recommended that the extra work amounts to less than, for example, 5 h per week, if you have 45 h of study per week. However, it would be better to work only during summer vacation. If you have to work more alongside your studies, e.g., 15 h per week, then it is recommended to have a 5 ECTS course less than usual, which frees up 9–10 hours per week. However, you must be aware that there may be a requirement for a certain progress in the educational process, e.g. in the first year of study.

How extra work may affect the study

The study and additional work should not be summed up to 37 or 45 h per week. It is only the study, which should count 45 h per week! Additional work in addition to the study should be added to the 45 h per week. This can make you extra tired if you work 45 h a week with your studies. As a result, you learn more slowly and are less efficiently. If extra work causes the study to be less efficient or is reduced below 45 h per week, it can result in poor understanding, more difficulty understanding the lessons and solving the problems, and a reduced joy with the study. The result could be a failure or poor results in the exam. This could be critical for the remaining part of the education. The courses in electrical circuit theory and in electrophysics in the first year of electrical engineering education are extremely important as they form the basis for understanding the rest of the education! Therefore, it is important to understand these courses extremely well. If you drop out during the first year, it can extend your education. Maybe you have to start over one year later. If the purpose of the extra work is to save money for a vacation, one should consider what is most important. Perhaps one could try to adjust the budget to what is possible from educational support. Perhaps the state offers favourable loans especially for students as additional support. These loans can be interest-free while you are studying. After completing your education, you can easily repay such government loans. It might be a better solution than exhausting yourself with extra work while studying.

How to record how much you work

It could be recommended to write down every day in a spreadsheet how many hours you have worked on the different courses and the extra work in addition to the study. Thus, it is possible to get an overview of the amount of work used for the different courses, as well as the total work per week.

It is important to record how much you work on the various courses. Then you can see if you, e.g., reaches 9–10 hours per week for each 5 ECTS course. If this is difficult, then it is recommended (if possible) to leave the extra work along with the studies. Next, you can consider postponing a course until later in the education. Here, a course is chosen that is not strictly necessary for the next semester.

If you drop the extra work or postpone a course until later in the education, it is an advantage to do it early in the semester. This gives you more time to complete the remaining courses in the semester. If you wait too long, the course that you delay until later will have already taken a lot of time away from the other courses that you want to complete.

Ultimately, studying should be a pleasure in addition to being hard work, and it will only be a pleasure if you give yourself enough time to understand the material.

1.3 How to Organize the Electrical Circuit Theory Courses

1.3.1 How to Work with the First Two Electrical Circuit Theory Courses

The author has two 5 ECTS credit courses in the first semester of the power engineering programme: In a 13-week period, the first course with Chaps. 1–4 is completed over 6,5 weeks with two lessons per week (13 lessons in total). The second course with Chaps. 5–10 is then completed in the remaining 6,5 weeks (also 13 lessons in total). During the 13 weeks, there are, therefore, in total 26 lessons with 26 tests and 2–3 experimental exercises.

This volume is the first volume of this textbook, covering chapters 1–4 corresponding to the first course.

The second volume of this textbook covers chapters 5–10 corresponding to the second course.

The third volume of this textbook covers chapters 11–15 corresponding to the third course, which is given in the second semester in a period of 13-weeks with 13 lessons that include 13 tests and 2 experimental exercises.

Lessons The two classes at DTU are every Wednesday and Friday morning from 8 am to 12 pm. Each class is organised in this way:

- At 0800–0845: Written test without aids in the material covered in the previous lesson.
- At 0900–1200: Short lectures (20–30 minutes per lecture) combined with short periods where the students solve small problems in class (approx. 20 minutes per period).
- At 1300–1700: Some students choose to sit in small groups and prepare for the next lesson. Other students choose to prepare individually at home.

Homework consists of:

- Finish the problems that started in the morning. These problems can be solved with the help of this book and possibly while sitting together with other students. The problems are handed in later and will be registered and possibly corrected if there is sufficient assistant teacher capacity.
- Reading the material that was reviewed at today's lecture and looking at the notes you made in the morning at the lecture.
- Finally, prepare for the test that comes in the morning in the next lesson. You can read the guide questions for this test in advance. The idea is that you find the answers to the test's questions while reading the theory in this book. To start with, you may have to memorize a few formulas. When you have used the formulas a few times in the test and when solving the problems, you no longer have to make an effort to memorize them. You may have to remember a formula or two for a test. However, the vast majority of the test comes naturally as a logical progression or explanation, which is not difficult to remember because it is based on an understanding of the logical context of the material.

1.3.2 The Activity Plan for the First Electrical Circuit Theory Course

The lessons and the tests

All lessons are 4 h long, typically 8–12. The test starts at 8 am. The lecture combined with exercises takes place between 9 am and 12 pm (Table 1.1).

On the first day, you meet at 8 am and get the first test, which counts towards character development on an equal footing with the other tests. This motivates you to also meet on the first day, where after the test you get the first lecture, which describes how to work with the course and gives an overview of the course's academic content.

On the second day, you meet again at 8 am and get the second test, which deals with important elements from the lecture that was covered on the first day and described in Chap. 1.

Table 1.1 The activity plan for the first electrical circuit theory course

Lesson no.	Test number	Topics covered in the lecture
1	1.1	Chapter 1 General information. How to work with the course. Start with Chap. 2 (fractions and parentheses)
2	1.2, 2.1	Chapter 2 Introduction to mathematical topics (Exponential and power functions, differentiation)
3	2.2	Chapter 2 Introduction to mathematical topics (Integration, Linear equations with 2 unknowns)
4	2.3	Chap. 2 Introduction to mathematical topics (Example using the Node voltage method. The Supernode voltage method)
5	4.2	Chapter 2 Introduction to mathematical topics (Linear equations with 3 unknowns, example using the Mesh-Current method)
6	4.3	Chapter 3 Current, voltage and resistance
7	3.1	Chapter 3 ending. Chapter 4 Resistive circuits (Series and parallel connections, circuit simplification, Thévenin and Norton equivalent circuits and transformations)
8	4.1	Chapter 4 Resistive circuits (Superposition)
9	4.5	Chapter 4 Resistive circuits (Y-Delta transformation)
10	4.4	Chapter 4 ending
11	Exam	Company visit
12		Experimental exercise, Company visit
13		Experimental exercise, Company visit

Each of the 13 days, you meet at 8 am and take a test on what was covered on the previous day. The first day test has been conducted for good reasons. It is a test of general knowledge of very basic mathematics in primary and secondary school. This test can be an appetizer for the following two tests, which also deal with very elementary mathematics, which is an absolute necessity to be able to work with the course and associated assignments.

How to prepare the tests Preparation for the next day can be done in the afternoon, e.g. from 1 to 5 pm. You can see the guiding questions in the next test in advance. For example, you can read the first question in the next test and then read the text of the textbook until you find the passage in the textbook that gives the answer to the first question. Then you formulate this answer in your own words on a piece of paper.

Then read the next question and continue reading the text until you find the answer to this question and write the answer in your own words on the paper that becomes a draught for the next test. In this way, you work through all the questions. You get to read the text of the textbook in a more fun and motivated way than if you had to read the entire chapter from a to z without knowing what was important. The test questions are constructed in such a way that you get an overview of the most important elements of the chapter in question. When you have finished the draught for the next test in this way, you can go through it again just by looking at the questions and without looking at the text in the textbook. In this way, you prepare to take the test the next day without aids.

Typically, it takes about two hours to prepare for the next test.

Problems and oral exam The rest of the afternoon is typically spent completing problems that may have started in the morning. These problems may be handed in as homework the following day. If the course has an oral exam where you can pick one of the 13 tests, you may want to spend some time preparing to be able to present the test in question orally and explain what happens in the test. It is an advantage to repeat this while you can remember everything because of the written test you have prepared.

1.3.3 The Activity Plan for the Second Electrical Circuit Theory Course

One can work with the second course exactly the same way as described for the first course (Table 1.2).

1.3.4 Recommendations for the Teacher

Regarding the first lesson

As stated in the activity plan, there is a test on the very first day. This test serves two purposes:

1. It counts towards the final grade on an equal footing with the subsequent tests and motivates the students to meet at 8 am on the first day. It is an advantage for students and

Table 1.2 The activity plan for the second electrical circuit theory course.

Lesson no.	Test number	Topics covered in the lecture
1	2.3	Start with chapter no. 5 Capacitive circuits (Charging a capacitor connected to a voltage generator or to a current generator)
2	5.1	Chapter 5 Capacitive circuits (Charging a capacitor through a resistance)
3	5.2	Chapter 6 Inductive circuits (Charging a coil by a linear ramping current generator)
4	6.1	Chapter 6 Inductive circuits (Charging a coil through a resistance)
5	6.2	Chapter 7 s order circuits
6	7.1	Chapter 8 The Diode (Types of diodes. A diode in a series connection with a resistance. Diode rectifiers)
7	8.1	Chapter 9 The Transistor (The 3 states of a transistor. The IV output characteristics of a transistor. Transistor circuit with a collector resistance
8	9.1	Chapter 9 The Transistor (Depiction of I_C versus I_B. Depiction of V_{CE} versus I_B)
9	9.2	Chapter 10 The Operational Amplifier
10	10.1	Chapter 10 The Operational Amplifier
11	Exam	Company visit
12		Experimental exercise, Company visit
13		Experimental exercise, Company visit

for the teacher that everyone in the class comes along from the start and is introduced to how to work with the course and gets an overview of the course.

2. It can be a wake-up call for students. Students who start the programme can come from many different educational institutions and have very different knowledge of mathematics. This is not a criticism of the students. After all, the students' mathematics knowledge is a result of the teaching they received before they came to university! Students may discover different mathematical topics that they would like to brush up on. This motivates them for the next two lessons, which review some of the very fundamental mathematical topics that are important for understanding the theory and calculating problems in the subsequent teaching of electrical circuit theory.

The general assessment of one course

It is of course up to the course responsible or teacher how to determine the grades for the students' efforts. The author of this book presents the following grading scale to the students on the first day.

40% Tests (without aids in class)
40% Written or oral exam
10% One or two experimental exercises
10% Homework problems (with aids)

DTU uses a 7-step scale approximately as follows (grades are in parentheses): 0–30% (−3), 30–50% (00), 50–60% (02), 60–70% (4), 70–84% (7), 84–92% (10), 92–100% (12)

Written or oral exams are evaluated in the usual way.

Assessment of a test

The tests are assessed on a scale of 0–100%. A maximum of about 80% can be achieved if all formulas and calculations are correct. A higher assessment will depend on whether the following is met: All symbols must be explained the first time they appear. There must be an appropriate explanation accompanying the individual steps in the calculations and proofs. There must be a good structure with indentations and underlines, i.e. good order. The writing must be clear and easy to read. Small sketches are supplemented where this is an advantage or where it is required. Different colours may be used in the sketches.

How many tests are corrected?

The ideal is, of course, to correct all tests from all students. But it is seldom realistic to get enough teaching assistant hours, so the author uses the following method: When the class has completed 3 tests (i.e. after 3 lessons), a student rolls a die, and in this way chooses which of the three tests will be sent for correcting. The other two tests are not corrected in detail but are registered as completed if they are assessed as having been passed (better than about 50%).

If 13 tests are carried out, this means that, for example, 4 tests are corrected in detail with feedback and that the remaining 9 tests are registered as having been completed and passed. Neither the students nor the teacher know in advance which 4 tests are marked and which 9 tests are simply registered as completed and passed. It may be decided that the first test on the first day will be one of the 9 tests that are simply registered as completed and passed. Students should be allowed to try it out in class and on the first day they have not received prior instruction.

The final grade for the tests

The author calculates the final grade for the tests in this way:

Students are given the opportunity to view a table in which each student can see the grades for all the tests completed. The names of the students have been replaced with an alias number to anonymise the students. For example, on the first day, the students have anonymously chosen a four-digit alias number that the teacher knows. The students can then contact the teacher if there are tests that were completed but have not been registered or there are tests where the assistant teacher has graded in a way that the student does not understand. Any errors in the registration can be corrected by the teacher before the final grade is given. By involving students in this way, subsequent complaints are minimized.

When all registrations have been verified with the students, the final grade can be determined as follows: The average grade in % of the 4 tests that were corrected is multiplied by a factor that is the number of tests taken divided by the maximum possible number of tests. For example, if a student has obtained an average of 90% in the 4 tests, but has missed one of the other tests, then the student has only taken 12 tests, and the grade will be 90% x 12/13 = 83%. Students may be given the opportunity to take one missing test at the end of the course if absence is due to illness.

The experimental exercise is typically carried out in small groups of 2–3 students. Of course, a grade can be assigned to the submitted report. Alternatively, the instructor can read the report, give feedback, and give a pass grade. For example, one can give each student 3% for participating in the experimental exercise, 3% for participating in the report writing, and an additional 4% for the approved report, i.e. 10% in total.

Assessment of homework problems

The ideal is, of course, to correct all homework problems for all students. But it is rarely realistic to get enough teaching assistant hours, so one may use the following method.

A sample is taken in which the assistant teacher corrects a homework assignment. The 10% are graded according to the quality of the corrected homework assignment and the number of homework assignments submitted, i.e., approximately the same system as with tests. When homework assignments are returned after approval, the teacher or assistant teacher can review the homework assignments on the board. Then everyone in the class can correct their assignments themselves.

Lesson no. 11,12,13

Exams, company visits, and experimental exercises are included in the schedule as lessons 11,12,13. Of course, it is up to the teacher to include these elements in the appropriate place in the course. The three lessons can also be used for other theoretical or practical topics.

1.4 Tests

The tests are without any aids; only pencils, rulers, etc. are permitted.

1.4.1 Test 1.1 Basic Mathematics

Problem 1 (25%) Examples of fractions
Let $a > 0$ *and* $b > 0$ and consider the fraction $\frac{a}{b}$.
A proper fraction has $a < b$. An improper fraction has $a \geq b$.
NB! If the result turns out to be an improper fraction, the result should also be presented as a mixed fraction , i.e. an integer number plus a proper fraction.

A. Multiplication of fractions $\frac{2}{7} \cdot \frac{16}{3} =$
B. Division of fractions $\frac{3}{8} : \frac{9}{5} =$
C. Sum of fractions $\frac{9}{7} + \frac{2}{5} =$

Problem 2 (20%) Please solve the following problems in brackets.

A. $a(b + c) =$
B. $(a + b)(c + d) =$
C. $(a - b + c)(d + e) =$
D. $(a + b)(a + b) = (a + b)^2 =$
E. $(a + b)(a - b) =$

Problem 3 (12%) Complete the following rules for exponential functions.
Let $a > 0$ and let x and y be variables.

A. $a^x a^y =$
B. $\dfrac{a^x}{a^y} =$
C. $(a^x)^y =$

Problem 4 (10%) Please calculate the following fraction.

$$\frac{(10^2)^4 \cdot (10^2)^{1/2}}{(100^3)^{1/2}}$$

Problem 5 (8%) Please complete the following rules for power functions.
Let a be a constant real number, and let x and y be variables.

A. $x^a y^a =$
B. $\dfrac{x^a}{y^a} =$

Problem 6 (8%) Please differentiate (↷) the following functions.

A. $8\,cos(5x)$ ↷
B. $4\,(e^x + e^{3x})$ ↷

Problem 7 (17%) Please calculate the following integrals.

A. $\int \frac{1}{2} e^{2x} dx$
B. $\int_0^4 \frac{1}{2} e^{2x} dx$

1.4.2 Test 1.2 Diary of Time Spent on Courses and the Use of Laptop During Lectures

Problem 1 (40%)

A. How many hours is a study year considered?
B. How many weeks is a study year considered?
C. How many study hours per week?
D. How many hours in total is a 5 ECTS point course considered to be?

Problem 2 (20%)
Why would it be a good idea to keep a diary of time spent on each course and extra work in a spreadsheet?

Problem 3 (40%)

Please quote four reasons not to use laptops during a lecture:

1. The first reason?
2. The second reason?
3. The third reason?
4. The fourth reason?

You can use the following abbreviations:

A students do not use the laptop during the lesson
B students use a laptop only to take notes
C students use laptops for other purposes during the lecture

Mathematical Foundation

2

In this chapter, we will introduce the mathematical topics used in electrical circuit theory. As first semester students have a variety of educational backgrounds, there is naturally a variety of mathematical experience in the class. The purpose of this chapter is for everyone to review the necessary mathematical subjects from the beginning to an extent sufficient to complete the first semester in circuit theory.

2.1 Fractions

Proper and improper fractions

Consider the fraction

$$\frac{a}{b} \quad where\ a\ and\ b\ are\ integers \tag{2.1}$$

$$A\ proper\ fraction\ has\ a < b \tag{2.2}$$

$$An\ improper\ fraction\ has\ a \geq b \tag{2.3}$$

If a result turns out to be improper fraction, the result can also be presented as a mixed fraction, which is an integer plus a proper fraction.

EXAMPLE OF A CONVERSION FROM AN IMPROPER FRACTION TO A MIXED FRACTION

$$\frac{26}{5} = \frac{25+1}{5} = \frac{25}{5} + \frac{1}{5} = 5 + \frac{1}{5}$$

N. Dam, *An Introduction to Electrical Circuit Theory*, Synthesis Lectures on Engineering, Science, and Technology, https://doi.org/10.1007/978-3-032-34346-8_2

EXAMPLE OF A CONVERSION FROM A MIXED FRACTION TO AN IMPROPER FRACTION

$$7+\frac{1}{7}=\frac{49}{7}+\frac{1}{7}=\frac{49+1}{7}=\frac{50}{7}$$

In the following, *a, b, c, d* are real numbers.

Sum or difference of fractions

$$\frac{a}{b}\pm\frac{c}{d}=\frac{a}{b}\cdot\frac{d}{d}\pm\frac{c}{d}\cdot\frac{b}{b}=\frac{ad\pm cb}{bd} \tag{2.4}$$

EXAMPLE

$$\frac{2}{7}+\frac{7}{3}=\frac{2}{7}\cdot\frac{3}{3}+\frac{7}{3}\cdot\frac{7}{7}=\frac{2\cdot 3+7\cdot 7}{7\cdot 3}=\frac{6+49}{21}=\frac{55}{21}=\frac{42+13}{21}=\frac{42}{21}+\frac{13}{21}=2+\frac{13}{21}$$

Multiplication of fractions

$$\frac{a}{b}\frac{c}{d}=\frac{ac}{bd} \tag{2.5}$$

EXAMPLE

$$\frac{2}{14}\cdot\frac{28}{3}=\frac{2}{3}\cdot\frac{28}{14}=2\frac{2}{3}=\frac{4}{3}=1+\frac{1}{3}$$

Division of fractions

$$\frac{\left(\frac{a}{b}\right)}{\left(\frac{c}{d}\right)}=\frac{a}{b}\cdot\frac{d}{c}=\frac{ad}{bc} \tag{2.6}$$

EXAMPLE

$$\frac{\left(\frac{3}{4}\right)}{\left(\frac{7}{2}\right)}=\frac{3}{4}\cdot\frac{2}{7}=\frac{6}{28}=\frac{3}{14}$$

OTHER EXAMPLES[1]

$$\frac{32000+400}{1600}=\frac{320+4}{16}=\frac{320}{16}+\frac{4}{16}=20+\frac{1}{4}=20{,}25$$

$$\frac{32000\cdot 400}{1600}=\frac{320\cdot 100\cdot 400}{1600}=\frac{320\cdot 400}{16}=\frac{320}{16}\cdot 400=20\cdot 400=8\ 000$$

2.2 Exponential Functions

Exponential functions are very common within circuit theory. An *exponential function* is a real function, which in its most basic form is written as follows.

[1] It should be noted that this book was written in an EU country (Denmark) where the comma is used as the decimal separator and the space is used as the thousands separator, e.g. 1 222,25. This differs from the American standard, where the example could be written as 1,222.25.

$$a^x \tag{2.7}$$

Here, the *exponent* x is the independent real variable, while base a is a real positive constant not equal to 1.

The most common functions are when $a = 10$ or when $a = e$:

$$10^x \tag{2.8}$$

or

$$e^x \tag{2.9}$$

Some important rules:

$$a^x a^y = a^{x+y} \tag{2.10}$$

$$\frac{a^x}{a^y} = a^{x-y} \tag{2.11}$$

$$(a^x)^y = a^{xy} \tag{2.12}$$

EXAMPLES

$$\frac{32000 \cdot 400}{1600} = \frac{32 \cdot 10^3 \cdot 4 \cdot 10^2}{16 \cdot 10^2} = \frac{32 \cdot 4}{16} \cdot 10^{3+2-2} = 2 \cdot 4 \cdot 10^3 = 8 \cdot 10^3 = 8\,000$$

$$\frac{(10^2)^3 \cdot (10^2)^{\frac{1}{2}}}{(100^3)^{\frac{1}{2}}} = \frac{10^{2 \cdot 3} \cdot 10^{2 \cdot \frac{1}{2}}}{(100^{\frac{1}{2}})^3} = \frac{10^6 \cdot 10^1}{10^3} = 10^{6+1-3} = 10^4$$

In the last equation, we use $100^{\frac{1}{2}} = \sqrt{100} = 10$

2.3 Power Functions

A power function is a real function which in its most basic form is written as follows.

$$x^a \tag{2.13}$$

Here, a is a real constant and x is a real positive variable[2].

An example is the well-known function x^2.

Important rules, where a and b are constants, while x and y are variables:

$$x^a\, x^b = x^{a+b} \tag{2.14}$$

$$\frac{x^a}{x^b} = x^{a-b} \tag{2.15}$$

[2] If a is a general real number (that is, a real number, which is not an integer), then x should be a positive variable.

$$(x^a)^b = x^{ab} \tag{2.16}$$

$$x^a \ y^a = (xy)^a \tag{2.17}$$

$$\frac{x^a}{y^a} = (\frac{x}{y})^a \tag{2.18}$$

Note that 2.14 is analogous to 2.10.
Likewise, Eq. 2.15 is analogous to Eq. 2.11.
Finally, 2.16 is analogous to 2.12.

EXAMPLES

$$2^{\frac{1}{2}} \cdot 50^{\frac{1}{2}} = (2 \cdot 50)^{\frac{1}{2}} = 100^{\frac{1}{2}} = 10$$

$$\frac{24^{\frac{1}{2}}}{6^{\frac{1}{2}}} = (\frac{24}{6})^{\frac{1}{2}} = 4^{\frac{1}{2}} = 2$$

$$\frac{(2500 + 1100)^{\frac{1}{2}}}{(260 + 140)^{\frac{1}{2}}} = (\frac{2500 + 1100}{260 + 140})^{\frac{1}{2}} = (\frac{3600}{400})^{\frac{1}{2}} = (\frac{36}{4})^{\frac{1}{2}} = 9^{\frac{1}{2}} = 3$$

The reader is encouraged to give small examples for the calculation rules 2.14, 2.15, and 2.16.

2.4 Differentiation

The fundamental definition

Figure 2.1 shows the definition of the differential quotient. Let x_o be a fixed point on the horizontal x axis, and let x be a variable point on the same axis. The slope of the blue curve secant through A and B is the increase Δf of the function f in the vertical y direction divided by the increase Δx of the first variable x in the x direction, which is the same as the increase in f divided by the increase in x :

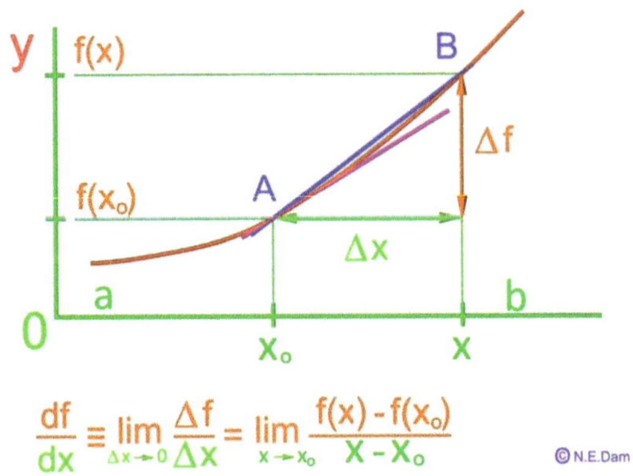

Fig. 2.1 Definition of the differential quotient

$$\text{The slope of secant AB} = \frac{\Delta f}{\Delta x} = \frac{f(x) - f(x_o)}{x - x_o} \tag{2.19}$$

If $\Delta x \to 0$, point B approaches point A and the secant AB approaches the purple tangent at point A. The slope of the secant AB, which is given by Eq. 2.19, then approaches the slope of the tangent at point A. It is denoted as the differential quotient at point A or the differential quotient of f at point x_o and is defined by the following.

$$\frac{df}{dx}(x_o) = \lim_{\Delta x \to 0} \frac{\Delta f}{\Delta x} = \lim_{x \to x_o} \frac{f(x) - f(x_o)}{x - x_o} \tag{2.20}$$

$\lim\limits_{x \to x_o}$ means the limit when x approaches x_o. The differential quotient $\frac{df}{dx}(x_o)$ is thus the limit of the secant slope $\frac{\Delta f}{\Delta x}$ when x approaches x_o. This limit is the tangent slope of the function f at the point x_o. The differential quotient can also be denoted in this way:

$$\frac{df}{dx}(x_o) = f'(x_o) \tag{2.21}$$

Conditions for the function $f(x)$ to be differentiable at the point x_o

- If the function $f(x)$ is to be differentiable at the point x_o, then the limit in Eq. 2.20 must be the same, regardless of whether x approaches x_o from the left or from the right side.
- As a consequence, $f(x)$ must be continuous in x_o.
- Furthermore, the tangent slope has a well-defined value at the point x_o, that is, the tangent should not be vertical.
- If f is differentiable at all points x_o in the interval between a and b, then f is said to be differentiable in the interval from a to b.

Differentiation rules for functions in general

We consider the functions $f(x)$ and $g(x)$. We assume that the functions are differentiable in the intervals considered. The arrow $\curvearrowright$ means the differentiation process:

Differentiation of a function:

$$f(x) \curvearrowright f'(x) = \frac{df}{dx} \tag{2.22}$$

Differentiation of a constant :

$$c \curvearrowright 0 \tag{2.23}$$

Example: $5 \curvearrowright 0$

Differentiation of a constant c times a function :

$$c\ f(x) \curvearrowright c\ f'(x) = c\ \frac{df}{dx} \tag{2.24}$$

Example: $5 \cdot x^2 \curvearrowright 5 \cdot 2x = 10x$

Differentiation of a sum or difference of two functions f(x) and g(x)

$$f(x) \pm g(x) \curvearrowright f'(x) \pm g'(x) = \frac{df}{dx} \pm \frac{dg}{dx} \tag{2.25}$$

Example: $5x^2 + 4x \curvearrowright 10x + 4$

Differentiation of a product of two functions $f(x)$ and $g(x)$

$$f(x) \cdot g(x) \curvearrowright f'(x) \cdot g(x) + f(x) \cdot g'(x) \tag{2.26}$$

Example: $5x^2 \cdot 4x \curvearrowright 10x \cdot 4x + 5x^2 \cdot 4 = 60x^2$

Differentiation of a quotient of two functions $f(x)$ and $g(x)$

$$\frac{f(x)}{g(x)} \curvearrowright \frac{f'(x) \cdot g(x) - f(x) \cdot g'(x)}{[g(x)]^2} \tag{2.27}$$

EXAMPLE

$$\tfrac{5x^2}{4x} \curvearrowright \tfrac{10x \cdot 4x - 5x^2 \cdot 4}{[4x]^2} = \tfrac{20x^2}{4^2x^2} = \tfrac{20x^2}{16x^2} = \tfrac{5}{4} \text{ or easier: } \tfrac{5x^2}{4x} = \tfrac{5x}{4} \curvearrowright \tfrac{5}{4}$$

Differentiation of a composite function, the "chain rule"

$$f[g(x)] \curvearrowright f'[g(x)] \cdot g'(x) \tag{2.28}$$

EXAMPLE

$$sin[5x^2] \curvearrowright cos[5x^2] \cdot 10x$$

Regarding the last rule, we differentiate the "outer" function f and leave the "inner" function g unaltered. Then we multiply by the differentiated "inner" function g'.

Differentiation rules for some important functions

We consider some of the most important functions $f(x)$. We assume that the functions are differentiable in the intervals considered. A constant is denoted by c. Note that the arrow $\curvearrowright$ means the differentiation process:

Differentiation of a function :

$$f(x) \curvearrowright f'(x) = \frac{df}{dx} \tag{2.29}$$

Differentiation of a sine function of x

$$\sin x \curvearrowright \cos x \tag{2.30}$$

Differentiation of a cosine function of x

$$\cos x \curvearrowright -\sin x \tag{2.31}$$

Differentiation of a power of x :

$$x^n \curvearrowright n \cdot x^{n-1} \tag{2.32}$$

$$\text{Example: } x^3 \curvearrowright 3 \cdot x^2$$

Differentiation of the natural logarithm[3] of x :

$$ln(x) = log_e(x) \curvearrowright \frac{1}{x} \tag{2.33}$$

Differentiation of an exponential function of x

$$a^x \curvearrowright a^x \cdot \log_e(a) \ , \ a > 0 \tag{2.34}$$

$$\text{Example: } 10^x \curvearrowright 10^x \cdot \log_e(10)$$

Differentiation of a standard exponential function of x

$$e^x \curvearrowright e^x \cdot \log_e(e) = e^x \ \cdot \ 1 = e^x \tag{2.35}$$

[3] The **logarithm** is denoted $\log_b(x)$ and it is pronounced as "the logarithm of x to base b" or "the base-b logarithm of x" or simply "the log, base b, of x". The **base** of the natural logarithm is $e \simeq 2,718$. If the base $b = e$, we have the natural logarithm $\log_e(x)$, which is sometimes also written $ln(x)$. For all logarithm functions we have $\log_b(b) = 1$, especially $\log_e(e) = 1$.

2.5 Half-Tangents

Most people are familiar with the concept of a tangent, which is a straight line that touches a curve at a point. At this point, the slope of the tangent is equal to the slope or differential quotient of the curve. One could say that the tangent is a straight-line approximation to the curve near the point. In connection with the charging and discharging of capacitors and coils, we will use the concept of half-tangents, which is therefore discussed in more detail below.

When $f(x)$ is discontinuous at x_o and $f'(x)$ is continuous at x_o

Figure 2.2 shows a function that is discontinuous at a point x_o. This means that we have to remove the pencil from the paper at x_o, when we draw the function on the paper. The function has different values on the left side and on the right side of x_o.

$$f(x) \to f(x_o-) \text{ when } x \to x_o- \text{ , i.e. } x \text{ approaches } x_0 \text{ from the left side} \tag{2.36}$$

$$f(x) \to f(x_o+) \text{ when } x \to x_o+ \text{ , i.e. } x \text{ approaches } x_0 \text{ from the right side} \tag{2.37}$$

$$f(x_o-) \neq f(x_o+) \tag{2.38}$$

If we want the function to be defined at x_o, then we can set a value for the function at this point; for example, we can determine that $f(x_o) = f(x_o+)$ as shown in Fig. 2.2.

Figure 2.2 also shows the half tangent from the left in x_o and the half tangent from the right in x_o. As seen in the figure, the two half-tangents have the same slope:

$$f'(x_o-) = f'(x_o+) \tag{2.39}$$

Note that $f'(x)$ is not defined for $x = x_o$. If we start from x_o, then the secant slopes on the right in x_o are well defined, while the secant slopes on the left in x_o approach infinity as x approaches x_o from the left.

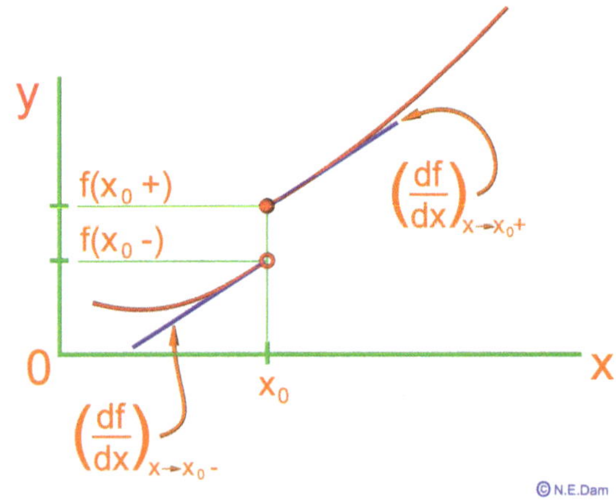

Fig. 2.2 Half-tangents for a discontinuous function at point x_o

As mentioned above under "Conditions for the function $f(x)$ to be differentiable at point x_o", $f(x)$ cannot be differentiable in x_o when $f(x)$ is discontinuous in x_o. If Eq. 2.39 is satisfied, and furthermore,

$$f(x_o-) = f(x_o) = f(x_o+) \tag{2.40}$$

then f is continuous and differentiable at the point x_o.

A function, where $f(x)$ is continuous at x_o and f'(x) is discontinuous at x_o

Figure 2.3 shows a continuous function at a point x_o. This means that we do not have to lift the pencil from the paper at x_o, when we draw the function on the paper. The function has the same value from the left side and from the right side of x_o.

$$f(x) \to f(x_o-) \text{ when } x \to x_o- \text{ , i.e. } x \text{ approaches } x_0 \text{ from the left side} \tag{2.41}$$

$$f(x) \to f(x_o+) \text{ when } x \to x_o+ \text{ , i.e. } x \text{ approaches } x_0 \text{ from the right side} \tag{2.42}$$

$$f(x_o-) = f(x_o) = f(x_o+) \tag{2.43}$$

The figure also shows the half tangent from the left in x_o and the half tangent from the right in x_o. They are both defined, but with different slopes. In this case, the tangent at x_o is not defined. We have the following:

$$f'(x_o-) \neq f'(x_o+) \Rightarrow f'(x_o) \text{ is not defined} \tag{2.44}$$

The function $f'(x)$ is therefore discontinuous at the point x_o.

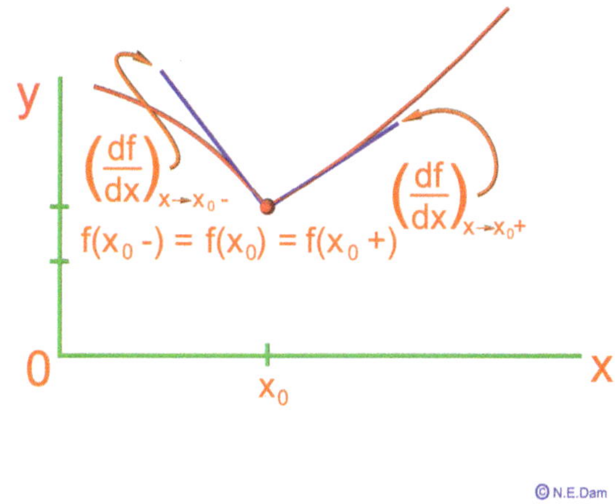

Fig. 2.3 Half-tangents for a continuous function at point x_o

2.6 Integration

How does the integral become the area between $f(x)$ and the x axis

Figure 2.4 shows the integration of a function $f(x)$. We consider a small interval of width dx around x shown as a rectangle. The height of this rectangle is $f(x)$. Hence, the area dA of the small rectangle is $f(x)$ times dx, that is, $dA = f(x)\,dx$.

Figure 2.5 shows the integration of the same function $f(x)$ from a to b. We consider many small intervals of width dx around x when x is increased in small steps from $x = a$ to $x = b$. The sum of all the small rectangles is the area A between the horizontal x axis $(y = 0)$ and the curve representing $f(x)$, at least in an approximation. When the width dx of the small rectangles is very small, this approximation becomes even better. In the limit, where $dx \to 0$, the sum of the many small areas becomes a perfect value for the area A.

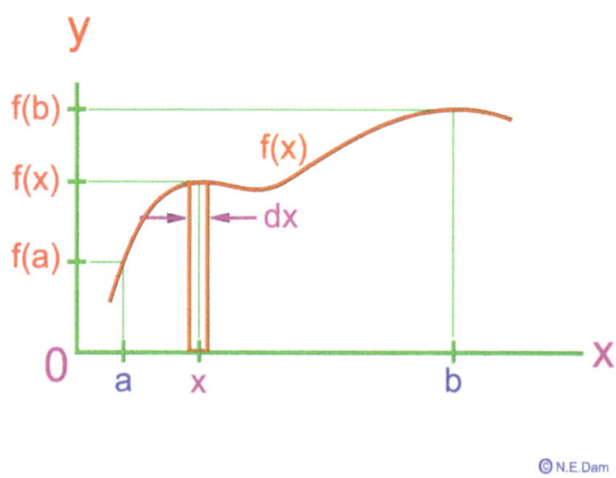

Fig. 2.4 Integration of a function $f(x)$ from a to b

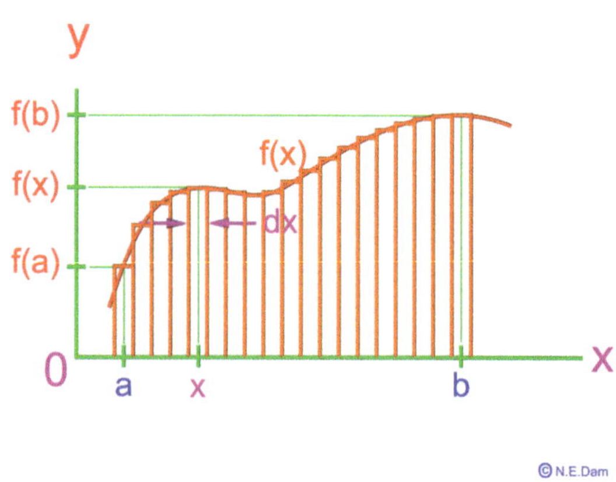

Fig. 2.5 Integration of a function $f(x)$ from a to b

This sum is called the integral of $f(x)$ from $x = a$ to $x = b$.

$$Area\ A = \int_{x=a}^{x=b} f(x)dx \tag{2.45}$$

Therefore, this integral determines the area between the horizontal x axis and the curve for $f(x)$ in the interval from a to b. In the case shown, the area lies above the x-axis and is counted as positive. If the area in a part of the interval from a to b lies below the x-axis, then this part of the area is counted as negative.

Requirements for the function $f(x)$

It is assumed that the function $f(x)$ is integrable, which in practice means that $f(x)$ must be bounded and piecewise continuous in the interval from a to b. The function $f(x)$ must therefore not go towards plus or minus infinity. If there is a point c between a and b, where $f(x)$ is not continuous, the interval from a to b can be divided into two intervals (a to c and c to b), where $f(x)$ is continuous in each of the two intervals, that is, $f(x)$ is piecewise continuous.

The integral in Eq. 2.45 is called **definite integral** as the integral is provided with limits (a and b) and therefore a specific value of the integral can be calculated. An integral without limits such as the integral in Eq. 2.46 is called an **indefinite integral**.

$$\int f(x)dx = F(x) + c\ ,\ \ c \in \Re \tag{2.46}$$

Note that $F(x)$ represents a single antiderivative of $f(x)$ with the property that if $F(x)$ is differentiated, we get the function $f(x)$ that is under the integral sign. When we add the constant c on the right-hand side of Eq. 2.46, it is because $F(x) + c$ is also differentiated into $f(x)$. That is, $F(x) + c$ represents all antiderivatives of $f(x)$ when c is an arbitrary (real) constant.

Determination of the integral

When we calculate the integral of $f(x)$, we find another function $F(x)$, which is the **antiderivative** of $f(x)$.

$$f(x) \rightleftharpoons F(x) + c = \int f(x)dx \tag{2.47}$$

$$\rightharpoonup \text{ is an integration process to the right} \tag{2.48}$$

$$\leftharpoondown \text{ is an differentiation process to the left} \tag{2.49}$$

The integration process is the opposite of a differentiation process. The task is therefore to find a function $F(x)$ for which it is true that if we differentiate $F(x)$, we get $f(x)$:

$$F'(x) = f(x) \leftharpoondown F(x) \text{ is the integration test} \tag{2.50}$$

There are different methods for calculating $F(x)$ corresponding to the different well-known functions. But instead of remembering all these integration rules, many may find it easier to just guess what $F(x)$ might be and then perform the integration test (2.50), i.e., differentiate $F(x)$ and see if it gives $f(x)$. We should always perform this integration test when we have calculated an integral!

We should always add a constant c to the antiderivative $F(x)$ if the integral is indefinite (without limits). This constant disappears if we differentiate $F(x) + c$. However, it is important to remember the constant since $F(x) + c$ should represent all functions with the property that, when differentiated, we get $f(x)$.

Example with $\int x^3 dx$
We wish to find the integral of x^3.

$$\int x^3 dx = F(x) + c \tag{2.51}$$

In other words, we should find a function $F(x)$ that, when differentiated, gives x^3. According to Eq. 2.32 on page 29 we guess that the desired function could be $F(x) = x^4$. Now, we perform the integration test (2.50) where we differentiate $F(x)$:

$$x^4 \curvearrowright 4x^3 \tag{2.52}$$

Note that the arrow $\curvearrowright$ means the differentiation process. We see that x^4 differentiates into $4x^3$. Hence, our guess was not quite correct. We modify our guess to be $F(x) = \frac{1}{4}x^4$. We perform the integration test (2.50) again and find:

$$\frac{1}{4}x^4 \curvearrowright \frac{1}{4} \cdot 4x^3 = x^3 \tag{2.53}$$

We are happy with the result because the integration test (2.50) ends with $f(x) = x^3$. However, there is a little more detail left. If we add a constant c to $F(x)$, we also end-up with x^3 , when we perform the integration test:

$$\frac{1}{4}x^4 + c \curvearrowright \frac{1}{4} \cdot 4x^3 + 0 = x^3 \tag{2.54}$$

Therefore, the complete solution is the following.

$$\int x^3 dx = F(x) + c = \frac{1}{4}x^4 + c \tag{2.55}$$

This is in fact a *family* of functions. If we change the constant c to another value, we get the same function $\frac{1}{4}x^4$ shifted up or down in the vertical direction. The family consists of parallel functions $\frac{1}{4}x^4 + c$ with different values of c.

EXAMPLE WITH $\int_0^3 x^3 dx$

We wish to find the integral from 0 to 3 of x^3.

$$\int_0^3 x^3 dx = [F(x) + c]_0^3 \tag{2.56}$$

Please, remember that we should use square brackets around $F(x) + c$ when we calculate the increase of $F(x) + c$. In the above example, we have already found $F(x) + c$ in the Eq. 2.55. We find the following

$$[F(x) + c]_0^3 = [\frac{1}{4}x^4 + c]_0^3 = [\frac{1}{4}3^4 + c] - [\frac{1}{4}0^4 + c] = \frac{1}{4}3^4 + c - c = \frac{81}{4} = \frac{80+1}{4} = \frac{80}{4} + \frac{1}{4} = 20{,}25 \tag{2.57}$$

Note that when we integrate with limits, we do not need the constant c, because it disappears in the calculation (c minus c). We can just write

$$\int_0^3 x^3 dx = [F(x)]_0^3 \tag{2.58}$$

2.7 Two Equations in Two Unknowns

A systematic description of solving linear equations using matrices and determinants requires a 5-ECTS point course. This is not included in this introduction. For the present, this has not been introduced before the second semester.

The following method used to solve two equations in two unknowns is called *Cramer's method*. Cramer's method may seem a bit complicated at first, but once you have tried it, it is easy to remember because it is very systematic. The method also has the advantage that the same procedure can be used for 3 or more equations with 3 or more unknowns.

We consider the following two linear equations:

$$\begin{aligned} a_{11}x_1 + a_{12}x_2 &= b_1 \\ a_{21}x_1 + a_{22}x_2 &= b_2 \end{aligned} \tag{2.59}$$

If we use *matrices*, these two equations can be written as follows.

$$\begin{Bmatrix} a_{11} & a_{12} \\ a_{21} & a_{22} \end{Bmatrix} \begin{Bmatrix} x_1 \\ x_2 \end{Bmatrix} = \begin{Bmatrix} b_1 \\ b_2 \end{Bmatrix} \tag{2.60}$$

This *matrix equation* may be written in a very concentrated form

$$\mathbf{A}\,\mathbf{X} = \mathbf{B} \tag{2.61}$$

Here

$$\mathbf{A} = \begin{Bmatrix} a_{11} & a_{12} \\ a_{21} & a_{22} \end{Bmatrix} \tag{2.62}$$

$$\mathbf{X} = \begin{Bmatrix} x_1 \\ x_2 \end{Bmatrix} \tag{2.63}$$

$$\mathbf{B} = \begin{Bmatrix} b_1 \\ b_2 \end{Bmatrix} \tag{2.64}$$

Matrix **A** is called a matrix of 2 by 2 or a matrix of 2nd order. In general, matrices are written in capital letters in bold. If you write on paper, it is difficult to write in bold. Instead, you can write a matrix as a capital letter with two dashes above it, e.g. $\bar{\bar{A}}$.

Index 11 (which should be pronounced one-one and not eleven) in element a_{11} means the first row and the first column. This index means that a_{11} is in the first row and in the first column. Similarly, the index 21 in element a_{21} means that a_{21} is located in the second row and in the first column.

Now, we form the *determinant* of matrix **A**

$$A = |\mathbf{A}| = \begin{vmatrix} a_{11} & a_{12} \\ a_{21} & a_{22} \end{vmatrix} = a_{11}a_{22} - a_{21}a_{12} \tag{2.65}$$

Furthermore, we also form the following two determinants, where we replace the first and second columns, respectively, by the elements of matrix **B**.

$$A_1 = \begin{vmatrix} b_1 & a_{12} \\ b_2 & a_{22} \end{vmatrix} = b_1a_{22} - b_2a_{12} \tag{2.66}$$

$$A_2 = \begin{vmatrix} a_{11} & b_1 \\ a_{21} & b_2 \end{vmatrix} = a_{11}b_2 - a_{21}b_1 \tag{2.67}$$

If $A = 0$, we have two parallel straight lines. If the two straight lines are coincident, then they have all points in common. If they are not coincident, but just parallel, they have no points in common. Hence, for $A = 0$, we have no intersection point in the usual meaning. In the following, we therefore assume $A \neq 0$, so we have an intersection point in the usual sense. The solution for the intersection points is

$$x_1 = \frac{A_1}{A} \text{ and } x_2 = \frac{A_2}{A} \tag{2.68}$$

Variables are generally denoted by x_1 and x_2. It depends on the circumstances, how they are denoted. It may be x and y, it may also be V_2 and V_3 as in the example below.

Perhaps you are wondering why we have to solve two equations with two unknowns in this apparently circumstantial way. Many will probably think that the method from primary school with two equations is simpler. But if we have three equations with three unknowns, it becomes difficult to use the method from primary school, and there is a great risk of error. If there are even more equations, then the primary school method is unmanageable. Cramer's method with matrices and determinants, on the other hand, is much more systematic. In principle, it can be used no matter how many equations there are. The reason why it is important to understand Cramer's method with only two equations with two unknowns is that by doing so we learn the procedure while it is very simple. Once we have learnt Cramer's method with two equations, it is easier to understand Cramer's method with three or more equations.

However, if we have 4 or more equations, it is easier to use the Gauss-Jordan elimination method, which is not included in this introduction.

2.8 Example Using the Node Voltage Method

This section shows how to use Cramer's method to find two unknown voltages at the nodes of the circuit in Figure 4.12. The circuit details are not shown in this section. We do not see how to set up the two equations with the two unknowns using the node voltage method. This is shown in Chapter 4 with Resistive Circuits. The example is designed to give two equations in two unknowns. We can thus use the procedure with two unknowns as shown in Sect. 2.7. Here and now we only consider how we use Cramer's method to solve the two equations. We therefore start from the following two equations with the two unknowns V_2 and V_3 (Fig. 2.6):

$$\begin{aligned} \frac{V_2 - V_1}{R_2} + \frac{V_2 - V_3}{R_3} + \frac{V_2 - V_4}{R_4} &= 0 \\ \frac{V_3 - V_1}{R_1} + \frac{V_3 - V_2}{R_3} + \frac{V_3 - V_4}{R_5} &= 0 \end{aligned} \tag{2.69}$$

The node potentials V_1 and V_4 are already known, they are $V_1 = V_{Gen}$ and $V_4 = 0$. In this example, we use the following very easy values for the resistances and for V_1:

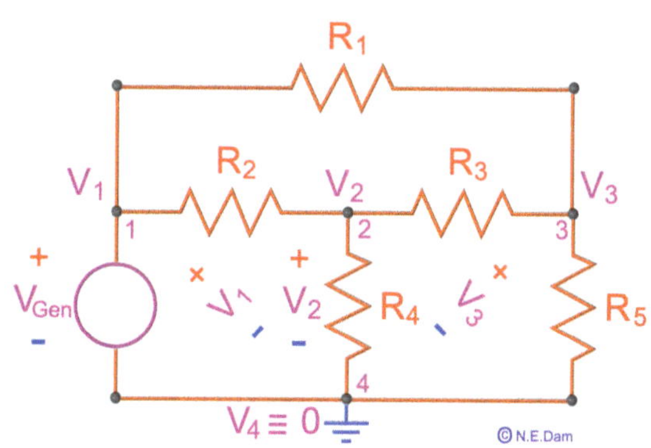

Fig. 2.6 Solving a circuit with two unknown voltages using Cramer's method

$$V_1 = 10\text{ V},\ R_1 = 10\ \Omega,\ R_2 = 10\ \Omega,\ R_3 = 10\ \Omega,\ R_4 = 10\ \Omega,\ R_5 = 10\ \Omega$$

Since all resistance values in the denominators in Eqs. 2.69 are equal to 10 Ω, we can multiply the two equations by 10 Ω on both sides of the equal signs. We find the following

$$\begin{aligned} 3V_2 - 10 - V_3 &= 0 \\ 3V_3 - 10 - V_2 &= 0 \end{aligned} \tag{2.70}$$

Now, we let $x_1 = V_2$ and $x_2 = V_3$ and rearrange the equations:

$$\begin{aligned} 3x_1 - x_2 &= 10 \\ 3x_2 - x_1 &= 10 \end{aligned} \tag{2.71}$$

Finally, we rearrange the equations so that x_1 in the upper equation is above x_1 in the lower equation and x_2 in the upper equation is above x_2 in the lower equation:

$$\begin{aligned} 3x_1 - x_2 &= 10 \\ -x_1 + 3x_2 &= 10 \end{aligned} \tag{2.72}$$

These two equations are put into matrix form using Eq. 2.127:

$$\begin{Bmatrix} 3 & -1 \\ -1 & 3 \end{Bmatrix} \begin{Bmatrix} x_1 \\ x_2 \end{Bmatrix} = \begin{Bmatrix} 10 \\ 10 \end{Bmatrix} \tag{2.73}$$

We calculate the determinants using Eqs. 2.65, 2.66, and 2.67:

$$A = |\mathbf{A}| = \begin{vmatrix} 3 & -1 \\ -1 & 3 \end{vmatrix} = 3 \cdot 3 - (-1) \cdot (-1) = 8 \tag{2.74}$$

$$A_1 = \begin{vmatrix} 10 & -1 \\ 10 & 3 \end{vmatrix} = 10 \cdot 3 - (10) \cdot (-1) = 40 \tag{2.75}$$

$$A_2 = \begin{vmatrix} 3 & 10 \\ -1 & 10 \end{vmatrix} = 3 \cdot 10 - (-1) \cdot (10) = 40 \tag{2.76}$$

Finally, we use Eq. 2.68 and find

$$V_2 = x_1 = \frac{A_1}{A} = \frac{40}{8} = 5\text{ V and } V_3 = x_2 = \frac{A_2}{A} = \frac{40}{8} = 5\text{ V} \tag{2.77}$$

As long as we set up equations and calculate determinants, we only calculate "pure" numbers without units. We add the unit V for the Volts at the end in the final result.

2.9 Three Equations in Three Unknowns

Before reading this section, it is highly recommended to read and understand Sect. 2.7. As mentioned in the beginning of Sect. 2.7 it requires a 5 ECTS point course to give a systematic description in solving linear equations using matrices and determinants. For the present, this has not been introduced before the second semester. Right now, we state the solution without proof.

We consider the following three linear equations

$$\begin{aligned} a_{11}x_1 + a_{12}x_2 + a_{13}x_3 &= b_1 \\ a_{21}x_1 + a_{22}x_2 + a_{23}x_2 &= b_2 \\ a_{31}x_1 + a_{32}x_2 + a_{33}x_3 &= b_3 \end{aligned} \tag{2.78}$$

If we use *matrices*, these three equations can be written as follows.

$$\begin{Bmatrix} a_{11} & a_{12} & a_{13} \\ a_{21} & a_{22} & a_{23} \\ a_{31} & a_{32} & a_{33} \end{Bmatrix} \begin{Bmatrix} x_1 \\ x_2 \\ x_3 \end{Bmatrix} = \begin{Bmatrix} b_1 \\ b_2 \\ b_3 \end{Bmatrix} \tag{2.79}$$

Note, e.g. the upper equation in 2.78 appears as follows from Eq. 2.79. Take the scalar product between the top row of matrix **A** and the column of matrix **X**, and then set this scalar product equal to the top element of matrix **B**.

The *matrix equation* may be written in a concentrated form

$$\mathbf{A}\,\mathbf{X} = \mathbf{B} \tag{2.80}$$

Here

$$\mathbf{A} = \begin{Bmatrix} a_{11} & a_{12} & a_{13} \\ a_{21} & a_{22} & a_{23} \\ a_{31} & a_{32} & a_{33} \end{Bmatrix} \tag{2.81}$$

$$\mathbf{X} = \begin{Bmatrix} x_1 \\ x_2 \\ x_3 \end{Bmatrix} \tag{2.82}$$

$$\mathbf{B} = \begin{Bmatrix} b_1 \\ b_2 \\ b_3 \end{Bmatrix} \tag{2.83}$$

Matrix **A** is called a 3 by 3 matrix or a matrix of 3rd order. Now, we form the *determinant* of the matrix **A**

$$A = |\mathbf{A}| = \begin{vmatrix} a_{11} & a_{12} & a_{13} \\ a_{21} & a_{22} & a_{23} \\ a_{31} & a_{32} & a_{33} \end{vmatrix} \tag{2.84}$$

In the next section, we will see how to calculate a 3 by 3 determinant. Furthermore, we also form the following three determinants, where we replace the first, second, respectively, the third column in determinant A by the elements of matrix **B**.

$$A_1 = \begin{vmatrix} b_1 & a_{12} & a_{13} \\ b_2 & a_{22} & a_{23} \\ b_3 & a_{32} & a_{33} \end{vmatrix} \tag{2.85}$$

$$A_2 = \begin{vmatrix} a_{11} & b_1 & a_{13} \\ a_{21} & b_2 & a_{23} \\ a_{31} & b_3 & a_{33} \end{vmatrix} \tag{2.86}$$

$$A_3 = \begin{vmatrix} a_{11} & a_{12} & b_1 \\ a_{21} & a_{22} & b_2 \\ a_{31} & a_{32} & b_3 \end{vmatrix} \tag{2.87}$$

The three equations in 2.78 represent three planes.

If $A \neq 0$, the three planes intersect at exactly one point. The solution of the system of equations then gives the coordinates for the point where the three planes intersect.

$$x_1 = \frac{A_1}{A} \text{ and } x_2 = \frac{A_2}{A} \text{ and } x_3 = \frac{A_3}{A} \tag{2.88}$$

If $A = 0$ there will not be exactly one intersection point. Then there are the following options for the three planes:

- They are parallel
- They are coincident
- They intersect in a straight line
- They intersect in three parallel straight lines

How to multiply a matrix by a constant

Let c be a real or complex constant. Then

$$c\mathbf{A} = c \begin{Bmatrix} a_{11} & a_{12} & a_{13} \\ a_{21} & a_{22} & a_{23} \\ a_{31} & a_{32} & a_{33} \end{Bmatrix} = \begin{Bmatrix} ca_{11} & ca_{12} & ca_{13} \\ ca_{21} & ca_{22} & ca_{23} \\ ca_{31} & ca_{32} & ca_{33} \end{Bmatrix} \tag{2.89}$$

In other words, all elements of the matrix are multiplied by the constant.

How to multiply a determinant by a constant

Let c be a real or complex constant. Then

$$cA = c\begin{vmatrix} a_{11} & a_{12} & a_{13} \\ a_{21} & a_{22} & a_{23} \\ a_{31} & a_{32} & a_{33} \end{vmatrix} = \begin{vmatrix} ca_{11} & ca_{12} & ca_{13} \\ a_{21} & a_{22} & a_{23} \\ a_{31} & a_{32} & a_{33} \end{vmatrix} = \begin{vmatrix} a_{11} & a_{12} & a_{13} \\ ca_{21} & ca_{22} & ca_{23} \\ a_{31} & a_{32} & a_{33} \end{vmatrix} = \begin{vmatrix} a_{11} & a_{12} & a_{13} \\ a_{21} & a_{22} & a_{23} \\ ca_{31} & ca_{32} & ca_{33} \end{vmatrix} \tag{2.90}$$

In other words, all elements in one row are multiplied by the constant. Alternatively, we can multiply all elements in one column by the constant:

$$cA = c\begin{vmatrix} a_{11} & a_{12} & a_{13} \\ a_{21} & a_{22} & a_{23} \\ a_{31} & a_{32} & a_{33} \end{vmatrix} = \begin{vmatrix} ca_{11} & a_{12} & a_{13} \\ ca_{21} & a_{22} & a_{23} \\ ca_{31} & a_{32} & a_{33} \end{vmatrix} = \begin{vmatrix} a_{11} & ca_{12} & a_{13} \\ a_{21} & ca_{22} & a_{23} \\ a_{31} & ca_{32} & a_{33} \end{vmatrix} = \begin{vmatrix} a_{11} & a_{12} & ca_{13} \\ a_{21} & a_{22} & ca_{23} \\ a_{31} & a_{32} & ca_{33} \end{vmatrix} \tag{2.91}$$

2.10 How to Calculate a 3 by 3 Determinant

We consider the determinant A from 2.84 again and use this determinant to show how to calculate a 3 by 3 determinant

$$A = |\mathbf{A}| = \begin{vmatrix} a_{11} & a_{12} & a_{13} \\ a_{21} & a_{22} & a_{23} \\ a_{31} & a_{32} & a_{33} \end{vmatrix} \tag{2.92}$$

Such a 3 by 3 determinant is more complicated to calculate than a 2 by 2 determinant. But, luckily, there is a procedure that can be used for all determinants (also for determinants of higher order than 3). Below is shown how to calculate the above 3 by 3 determinant starting from the top row in the determinant and then break the determinant down into three 2 by 2 determinants.

If we "expand after the first row", we find (please notice the systematically given signs)

$$A = a_{11}(-1)^{(1+1)}\begin{vmatrix} a_{22} & a_{23} \\ a_{32} & a_{33} \end{vmatrix} + a_{12}(-1)^{(1+2)}\begin{vmatrix} a_{21} & a_{23} \\ a_{31} & a_{33} \end{vmatrix} + a_{13}(-1)^{(1+3)}\begin{vmatrix} a_{21} & a_{22} \\ a_{31} & a_{32} \end{vmatrix} \tag{2.93}$$

and consequently

$$A = a_{11}\begin{vmatrix} a_{22} & a_{23} \\ a_{32} & a_{33} \end{vmatrix} - a_{12}\begin{vmatrix} a_{21} & a_{23} \\ a_{31} & a_{33} \end{vmatrix} + a_{13}\begin{vmatrix} a_{21} & a_{22} \\ a_{31} & a_{32} \end{vmatrix} \tag{2.94}$$

This requires further explanation. The first term on the right side of 2.93 is used as an example to explain the calculation:

$$a_{11}(-1)^{(1+1)}\begin{vmatrix} a_{22} & a_{23} \\ a_{32} & a_{33} \end{vmatrix} \tag{2.95}$$

It consists of three factors:

1. The first factor is the element a_{11}
2. The second factor is sign factor $(-1)^{(1+1)}$, i.e., minus one raised to the sum of the two numbers in the index, in this case 1 + 1 = 2. The sign factor then becomes $(-1)^2 = 1$
3. The third factor is the determinant $\begin{vmatrix} a_{22} & a_{23} \\ a_{32} & a_{33} \end{vmatrix}$. It is the determinant we get if we remove the row and the column where a_{11} is located, in determinant A, see Eq. 2.92. This determinant is called the sub-determinant or the minor of the element a_{11} and is denoted M_{11}.

The element a_{rs} is located in the row number r and the column number s. In general, we find that the subdeterminant or minor M_{rs} of element a_{rs} is defined as the determinant we get if we remove the row number r and the column number s in determinant A.

If we multiply the sign factor by the *minor*, we get the co-factor A_{rs} of the element a_{rs}

$$A_{rs} = (-1)^{(r+s)} M_{rs} \tag{2.96}$$

Then, the determinant in 2.93 can be written in a short form:

$$A = a_{11}A_{11} + a_{12}A_{12} + a_{13}A_{13} \tag{2.97}$$

The determinant A is calculated here (expanded) from row number 1. The determinant A can also be calculated (expanded) from row number 2:

$$A = a_{21}A_{21} + a_{22}A_{22} + a_{23}A_{23} \tag{2.98}$$

Similarly, it is possible to calculate determinant A from row number 3:

$$A = a_{31}A_{31} + a_{32}A_{32} + a_{33}A_{33} \tag{2.99}$$

The determinant A can also be calculated from any one of the three columns, e.g. the third column as shown here:

$$A = a_{13}A_{13} + a_{23}A_{23} + a_{33}A_{33} \tag{2.100}$$

If, for example, there is a zero element in column 3, then it is advantageous to calculate determinant A from column 3. This reduces the number of subdeterminants or"minors" from 3 to 2. The following examples show how to calculate various 3 × 3 determinants.

EXAMPLE 1 Calculation from row number 1 of the following determinant:

$$A = \begin{vmatrix} 2 & -1 & -1 \\ -1 & 2 & 0 \\ -1 & 0 & 2 \end{vmatrix}$$

We find

$$A = a_{11}(-1)^{(1+1)}\begin{vmatrix} a_{22} & a_{23} \\ a_{32} & a_{33} \end{vmatrix} + a_{12}(-1)^{(1+2)}\begin{vmatrix} a_{21} & a_{23} \\ a_{31} & a_{33} \end{vmatrix} + a_{13}(-1)^{(1+3)}\begin{vmatrix} a_{21} & a_{22} \\ a_{31} & a_{32} \end{vmatrix}$$

then

$$A = 2\begin{vmatrix} 2 & 0 \\ 0 & 2 \end{vmatrix} - (-1)\begin{vmatrix} -1 & 0 \\ -1 & 2 \end{vmatrix} + (-1)\begin{vmatrix} -1 & 2 \\ -1 & 0 \end{vmatrix}$$

and

$$A = 2 \cdot 4 + (1) \cdot (-2) + (-1) \cdot 2 = 4$$

The final result becomes

$$A = \begin{vmatrix} 2 & -1 & -1 \\ -1 & 2 & 0 \\ -1 & 0 & 2 \end{vmatrix} = 4 \tag{2.101}$$

EXAMPLE 2 Calculation from column number 3 of the following determinant:

$$A_1 = \begin{vmatrix} 1 & -1 & -1 \\ -1 & 2 & 0 \\ 1 & 0 & 2 \end{vmatrix}$$

We find

$$A_1 = a_{13}(-1)^{(1+3)}\begin{vmatrix} a_{21} & a_{22} \\ a_{31} & a_{32} \end{vmatrix} + a_{23}(-1)^{(2+3)}\begin{vmatrix} a_{11} & a_{12} \\ a_{31} & a_{32} \end{vmatrix} + a_{33}(-1)^{(3+3)}\begin{vmatrix} a_{11} & a_{12} \\ a_{21} & a_{22} \end{vmatrix}$$

which means

$$A_1 = (-1)\begin{vmatrix} -1 & 2 \\ 1 & 0 \end{vmatrix} - (0)\begin{vmatrix} 1 & -1 \\ 1 & 0 \end{vmatrix} + (2)\begin{vmatrix} 1 & -1 \\ -1 & 2 \end{vmatrix}$$

and therefore

$$A_1 = (-1) \cdot (-2) + 0 + (2) \cdot 1 = 4$$

The final result becomes

$$A_1 = \begin{vmatrix} 1 & -1 & -1 \\ -1 & 2 & 0 \\ 1 & 0 & 2 \end{vmatrix} = 4 \tag{2.102}$$

EXERCISE 1 Calculate the following determinant from column number 3. Show all steps in the calculation and show how to get the final result.

$$A_2 = \begin{vmatrix} 2 & 1 & -1 \\ -1 & -1 & 0 \\ -1 & 1 & 2 \end{vmatrix} = 0 \tag{2.103}$$

EXERCISE 2 Calculate the following determinant from row number 3. Show all steps in the calculation and show how to get the final result.

$$A_3 = \begin{vmatrix} 2 & -1 & 1 \\ -1 & 2 & -1 \\ -1 & 0 & 1 \end{vmatrix} = 4 \tag{2.104}$$

2.11 Example Using the Mesh-Current Method

This section shows how to use Cramer's method to find three unknown mesh-currents in the circuit in Fig. 2.7. The circuit details are not shown in this section. We do not see how to set up the three equations with the three unknowns using the Mesh-Current method. This is shown in Chapter 4 Resistive Circuits, Section 4.11. The example is designed to give three equations in three unknowns. We can thus use the procedure with three unknowns as shown in Sect. 2.9. Here and now we only consider how we use Cramer's method to solve the three equations. We therefore start from the following three equations with the three unknowns $x_1 = I_1, x_2 = I_2, x_3 = I_3$:

$$\begin{aligned}(R_2 + R_3)I_1 - R_3I_2 - R_2I_3 &= V_{G1} \\ (-R_3)I_1 + (R_3 + R_4)I_2 + (0)I_3 &= -V_{G2} \\ (-R_2)I_1 + (0)I_2 + (R_1 + R_2)I_3 &= V_{G2}\end{aligned} \tag{2.105}$$

If we compare 2.78 with 2.79 we see, that the three equations can be written in matrix form:

$$\begin{Bmatrix} (R_2 + R_3) & -R_3 & -R_2 \\ -R_3 & (R_3 + R_4) & 0 \\ -R_2 & 0 & (R_1 + R_2) \end{Bmatrix} \begin{Bmatrix} I_1 \\ I_2 \\ I_3 \end{Bmatrix} = \begin{Bmatrix} V_{G1} \\ -V_{G2} \\ V_{G2} \end{Bmatrix} \tag{2.106}$$

In this example, we have the following

$$V_{G1} = V_{G2} = 10\text{ V},\ R_1 = 10\ \Omega,\ R_2 = 10\ \Omega,\ R_3 = 10\ \Omega,\ R_4 = 10\ \Omega$$

We then find the following

$$\begin{Bmatrix} 20 & -10 & -10 \\ -10 & 20 & 0 \\ -10 & 0 & 20 \end{Bmatrix} \begin{Bmatrix} I_1 \\ I_2 \\ I_3 \end{Bmatrix} = \begin{Bmatrix} 10 \\ -10 \\ 10 \end{Bmatrix} \tag{2.107}$$

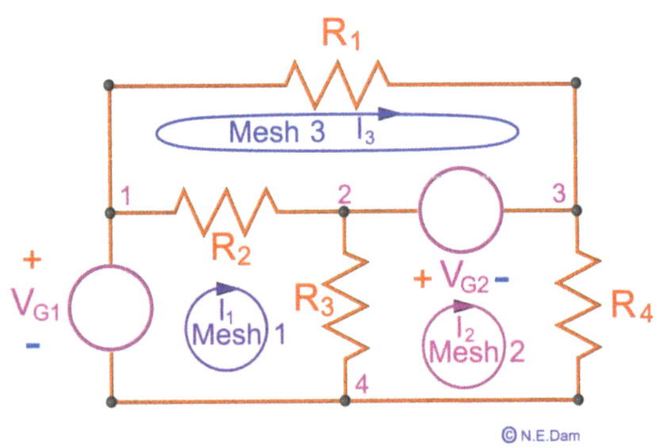

Fig. 2.7 Solving a circuit with three unknown mesh currents using Cramer's method

If we multiply by $\frac{1}{10}$ on both sides of 2.107 and use the rule for multiplication of a matrix by a constant from 2.89, we find the following

$$\frac{1}{10}\begin{Bmatrix} 20 & -10 & -10 \\ -10 & 20 & 0 \\ -10 & 0 & 20 \end{Bmatrix}\begin{Bmatrix} I_1 \\ I_2 \\ I_3 \end{Bmatrix} = \frac{1}{10}\begin{Bmatrix} 10 \\ -10 \\ 10 \end{Bmatrix} \tag{2.108}$$

$$\Rightarrow \begin{Bmatrix} 2 & -1 & -1 \\ -1 & 2 & 0 \\ -1 & 0 & 2 \end{Bmatrix}\begin{Bmatrix} I_1 \\ I_2 \\ I_3 \end{Bmatrix} = \begin{Bmatrix} 1 \\ -1 \\ 1 \end{Bmatrix} \tag{2.109}$$

It is much easier to calculate the determinants from Eq. 2.109 than from 2.107. We then use Cramer's method as described in Sect. 2.9. We write down the determinants A, A_1, A_2, A_3 using Eqs. 2.84, 2.85, 2.86, 2.87 and use the calculated results of 2.101, 2.102, 2.103, 2.104. We hereby find:

$$A = \begin{vmatrix} 2 & -1 & -1 \\ -1 & 2 & 0 \\ -1 & 0 & 2 \end{vmatrix} = 4$$

$$A_1 = \begin{vmatrix} 1 & -1 & -1 \\ -1 & 2 & 0 \\ 1 & 0 & 2 \end{vmatrix} = 4$$

$$A_2 = \begin{vmatrix} 2 & 1 & -1 \\ -1 & -1 & 0 \\ -1 & 1 & 2 \end{vmatrix} = 0$$

$$A_3 = \begin{vmatrix} 2 & -1 & 1 \\ -1 & 2 & -1 \\ -1 & 0 & 1 \end{vmatrix} = 4$$

According to 2.88 we find the final result

$$x_1 = I_1 = \frac{A_1}{A} = \frac{4}{4} = 1 \; and \; x_2 = I_2 = \frac{A_2}{A} = \frac{0}{4} = 0 \; and \; x_3 = I_3 = \frac{A_3}{A} = \frac{4}{4} = 1 \tag{2.110}$$

2.12 First-Order Differential Equation

Let $x(t)$ be a function with real value x of t. In the context of circuit theory, t is typically time. Now, we consider a first-order differential equation of the form

$$x'(t) + p(t)x(t) = q(t) \tag{2.111}$$

This can also be expressed as

$$\frac{dx}{dt} + p(t)x(t) = q(t) \tag{2.112}$$

As part of the solution, we have to calculate the integral

$$\int p(t)dt = P(t) + c \ , \ c \in \Re \tag{2.113}$$

This integral can be compared with Eq. 2.46 on page 33.

Note that $P(t)$ represents a single antiderivative of $p(t)$ with the property that if $P(t)$ is differentiated, we get the function $p(t)$ that is under the integral sign. When we add the constant c on the right-hand side of Eq. 2.113, it is because $P(t) + c$ is also differentiated into $p(t)$. That is, $P(t) + c$ represents all the antiderivatives of $p(t)$ when c is an arbitrary (real) constant.

The solution of the first order differential equation is then calculated this way:

$$x(t) = e^{-P(t)} \int e^{P(t)} q(t)dt \tag{2.114}$$

In Chap. 5 we will see how the first order differential equation is used to find the voltage across a capacitor as a function of time, when the capacitor is charged or discharged through a resistance. In Chap. 6 we will see how the first-order differential equation is used to find the current through a coil as a function of time, when the coil is 'charged' or 'discharged' through a resistance. In other words, first-order differential equations are used to analyse circuits consisting of a resistor combined with either a capacitor or an inductor.

2.13 Second-Order Differential Equation

In the following, we simply repeat the solutions of a second-order differential equation. In this introduction, a detailed description of the subject is not included.

Let $x(t)$ be a function with real value x of t. In the context of circuit theory, t is typically time. Now, we consider a second-order differential equation with constant coefficients of the form

$$a \cdot x''(t) + b \cdot x'(t) + c \cdot x(t) = f(t) \tag{2.115}$$

This is identical to

$$a \cdot \frac{d^2x}{dt^2} + b \cdot \frac{dx}{dt} + c \cdot x(t) = f(t) \tag{2.116}$$

We assume that $a, \ b$, and c are real constants and $f(t)$ is a real function of time. If $f(t) = 0$, then the differential equation is said to be a *homogeneous differential equation*, otherwise it is said to be a *non-homogeneous differential equation*. The procedure is to first find all solutions to the homogeneous differential equation. Then, you find one solution to the inhomogeneous equation. The total solution - also called the general solution - is then the solution to the homogeneous equation plus a solution to the inhomogeneous equation.

In Chap. 7 we will see how a second-order differential equation is used to find the voltage across a capacitor as a function of time, when the capacitor is in a series connection with a resistor and an inductor. In other words, second-order differential equations are used to analyse circuits consisting of a resistor combined with a capacitor and an inductor.

The homogeneous differential equation of second order

We consider a homogeneous differential equation of second order

$$a \cdot x''(t) + b \cdot x'(t) + c \cdot x(t) = 0 \tag{2.117}$$

We form the *characteristic equation*

$$a \cdot m^2 + b \cdot m + c = 0 \tag{2.118}$$

The roots are given by

$$m = \frac{-b \pm \sqrt{d}}{2a} \tag{2.119}$$

where the discriminant is

$$d = b^2 - 4ac \tag{2.120}$$

We have one of the following three cases for the roots:

1. Two real roots if d is positive
2. A repeated real root if d is zero
3. A pair of two conjugated complex roots if d is negative

The solutions $x_{Hom}(t)$ for the homogeneous second-order differential equation are listed below for the 3 cases:

1. Two real roots if d is positive

We have two different real roots $m_1 \neq m_2$. Then

$$x_{Hom}(t) = Ae^{m_1 t} + Be^{m_2 t},\ A, B \in \Re \text{ are real constants} \tag{2.121}$$

2. A repeated real root if d is zero

We have a repeated (or double) real root $m_1 = m_2 = m$. Then

$$x_{Hom}(t) = Ae^{mt} + B\ t\ e^{mt},\ A, B \in \Re \text{ are real constants} \tag{2.122}$$

3. A pair of two conjugated complex roots if d is negative
We have a pair of two conjugated complex roots

$$m_1 = \phi + j\omega \text{ and } m_2 = \phi - j\omega \text{ , } \phi, \omega \in \Re \text{ are real constants} \tag{2.123}$$

The solution is then

$$x_{Hom}(t) = e^{\phi t}\Big\{A\ cos(\omega t) + B\ sin(\omega t)\Big\}\ ,\ \ A, B \in \Re \text{ are real constants} \tag{2.124}$$

The imaginary unit is "i" defined by $i^2 = -1$. However, in circuit theory, we use the symbol "j" instead of "i" for the imaginary unit because "i" can be confused with a current. In principle, A and B could be complex constants, but if we are restricted to considering real solutions, A and B should be real constants.

Non-homogeneous second order differential equation
The general form of the non-homogeneous equation is as follows

$$a \cdot x''(t) + b \cdot x'(t) + c \cdot x(t) = f(t)\ ,\ \text{ where } f(t) \neq 0 \tag{2.125}$$

as stated in Eq. 2.115. The function $f(t)$ can be zero for certain values of t, but not for all values of t in the interval considered.

We find a solution $x_{Non}(t)$ that satisfies the nonhomogeneous second-order differential equation in Eq. 2.125. The total (general) solution is then

$$x_{Gen}(t) = x_{Hom}(t) + x_{Non}(t) \tag{2.126}$$

2.14 Tests

General guidelines for answering the test
The tests are without any aids; only pencils, rulers, etc. are permitted. The evaluation of the test, which is handed in, reflects not only the correctness of the answer but also the care and dedication that has been put into the answer.

A. Write the answer or formula correctly and show all the steps in the calculations.
B. The answer must be well structured, orderly, and easy to read.
C. Briefly explain the symbols used the first time they appear in the test.
D. If suggested, make a small sketch or diagram in colour to add clarity.

2.14.1 Test 2.1 Calculation with Fractions and Parentheses

NB! Points C and D in the above guidelines are not relevant in this test.

Problem 1 (25%) Rules for calculating with fractions

Let $a > 0$ *and* $b > 0$ and consider the fraction $\frac{a}{b}$.

A. What are a and b for defining a proper fraction?
B. What are a and b for defining an improper fraction?
C. Sum or difference of fractions $\frac{a}{b} \pm \frac{c}{d} =$
D. Multiplication of fractions $\frac{a}{b}\frac{c}{d} =$
E. Division of fractions $\frac{(\frac{a}{b})}{(\frac{c}{d})} = \frac{a}{b} : \frac{c}{d} =$

Problem 2 (32%) Examples of fractions

NB! If the result turns out to be an improper fraction, the result should also be presented as a mixed fraction , i.e. an integer number plus a proper fraction.

A. Multiplication of fractions $\frac{2}{7} \cdot \frac{14}{3} =$
B. Division of fractions $\frac{3}{8} : \frac{9}{4} =$
C. Sum of fractions $\frac{2}{7} + \frac{5}{3} =$
D. Conversion from a mixed fraction to an improper fraction $4 + \frac{1}{4} =$

Problem 3 (43%) Calculation with parentheses

Calculate the following problems with brackets:

A. (4%) $a(b + c) =$
B. (6%) $(a + b)(c + d) =$
C. (8%) $(a + b - c)(d + e) =$
D. (6%) the square of a two-term quantity $(a + b)(a + b) = (a + b)^2 =$
E. (6%) sum of two numbers times the difference of the same two numbers $(a + b)(a - b) =$
F. (13%) Simplify this fraction as much as possible $\frac{16a^2+16b^2+32ab}{4a^2-4b^2} =$

2.14.2 Test 2.2 Exponential and Power Functions, Fractions and Differentiation

Problem 1 (5%) Exponential function

Write a basic exponential function using a and x, explain a and x.

Problem 2 (9%) Rules for exponential functions

Complete the following rules for exponential functions

Let x and y be variables.

A. $a^x a^y =$

B. $\frac{a^x}{a^y} =$

C. $(a^x)^y =$

Problem 3 (18%) Calculations with fractions

Calculate the following fractions:
NB! If the result turns out to be an improper fraction, the result should also be presented as a mixed fraction , i.e. an integer plus a proper fraction.

A. $\frac{32000 \cdot 400}{1600} =$

B. $\frac{32000+400}{1600} =$

C. $\frac{(10^2)^3 \cdot (10^2)^{\frac{1}{2}}}{(100^3)^{\frac{1}{2}}} =$

Problem 4 (5%) Power function

Write a basic power function using a and x, explain a and x.

Problem 5 (6%) Rules for power functions

Let a be a constant real number, and let x and y be real positive variables.

A. $x^a y^a =$

B. $\frac{x^a}{y^a} =$

Problem 6 (16%) Calculations with fractions

Calculate the following fractions:

NB! If the result turns out to be an improper fraction, the result should also be presented as a mixed fraction , i.e. an integer number plus a proper fraction.

A. $\frac{24^{\frac{1}{2}}}{6^{\frac{1}{2}}} =$

B. $\frac{(2500+1100)^{\frac{1}{2}}}{(260+140)^{\frac{1}{2}}} =$

Problem 7 (41%) Differentiation

Let $\curvearrowright$ denote the differentiation process

A. (20%) Make a sketch (e.g., like in Fig. 2.1) showing the definition of the differential quotient of $y = f(x)$ at point x_o. Describe the conditions for the function f to be differentiable at the point x_o.
B. (3%) Differentiate $10\ cos(4x)$ $\curvearrowright$
C. (3%) Differentiate $5\ (e^x + e^{2x})$ $\curvearrowright$
D. (15%) Differentiate $\frac{5\ x^{\frac{1}{2}}}{10\ x^{\frac{1}{3}}}$ $\curvearrowright$

2.14.3 Test 2.3 Integration and Two Equations in Two Unknowns

Problem 1 (20%) Integration

A. (17%) Make a sketch (e.g., like in Fig. 2.5) showing the integral of $y = f(x)$ evaluated from a to b. Explain the formation of the definite integral relative to the sketch.
B. (3%) How do you check that the antiderivative function $F(x)$ is correct?

Problem 2 (24%) Calculation of integrals

Calculate the following integrals

A. $\int \frac{1}{4}e^{4x}dx$
B. $\int_0^4 \frac{1}{4}e^{4x}dx$
C. $\int 3e^{2x}dx$

Problem 3 (56%) Two equations in two unknowns

Intersection of straight lines

A. (12%) Make a sketch of the two lines: $y = 1x + 3$ and $y = 3x + 2$.
B. (8%) Show how to write the first equation in this form $1x_1 - 1x_2 = -3$ and do something similar for the second equation.
C. (8%) Write the two equations in matrix form

$$\begin{Bmatrix} a_{11} & a_{12} \\ a_{21} & a_{22} \end{Bmatrix} \begin{Bmatrix} x_1 \\ x_2 \end{Bmatrix} = \begin{Bmatrix} b_1 \\ b_2 \end{Bmatrix} \tag{2.127}$$

D. (12%) Please find the three determinats $A,\ A_1,\ A_2$ as shown in Eqs. 2.65, 2.66 and 2.67 on page 36 and page 36.
E. (8%) Please find the solution for the intersection point
$\{x_1\} =$
F. (8%) Equation $y = 3x + 2$ is now modified to $y = 1x + 2$. Make a sketch of the new line, write the matrix equation again, and show that $A = 0$. Could you make a connection between $A = 0$ and the relative orientation of the two lines?

Current, Voltage, and Resistance

3

The chapter opens with a definition of electric current I, electric voltage V, and electric resistance R, which are mutually related through the following current-voltage relationship

$$V = R \cdot I \tag{3.1}$$

This relation is called *Ohm's law*,[1] if the resistance is constant (i.e., independent of current at constant temperature).

Next, a section explains how to determine the sign of the current through a circuit element and the sign of the voltage across a circuit element. This is a configuration of a circuit element.

This is followed by a preliminary and short introduction to Kirchhoff's Current Law and Kirchhoff's voltage Law.[2] These two laws, together with Ohm's law, form the basis for all the methods used to calculate currents and voltages in pure resistive circuits.

The chapter ends with two sections that describe definitions and equivalent circuits for voltage sources[3] and current sources,[4] respectively.

---

[1] G. S. Ohm, Die galvanische Kette, mathematisch bearbeitet (Berlin, 1827), the title could be translated to "The Galvanic Circuit Investigated Mathematically".

[2] G. R. Kirchhoff, "Über die Auflösung der Gleichungen, auf welche man bei der Untersuchung der linearen Verteilung galvanischer Ströme geführt wird", published in Annalen der Physik und Chemie, vol. 72. The title could be translated to "On the solution of the equations to which one is led during the investigation of the linear distribution of galvanic currents".

[3] also known as a voltage generator V_{Gen}. Sometimes we consider a constant DC voltage source, which could be a battery V_{Bat}.

[4] also known as a current generator I_{Gen}.

N. Dam, *An Introduction to Electrical Circuit Theory*, Synthesis Lectures on Engineering, Science, and Technology, https://doi.org/10.1007/978-3-032-34346-8_3

3.1 Electric Current

An *electric current* is a charge flow, as shown in Fig. 3.1. In this figure, we consider an amount of charge dq, which passes through a cross-sectional area A. If we divide the amount of charge dq by the time dt it takes for the charge to pass through the area, we find the current.

Theorem 1 ***The definition of electric current*** *is the amount of charge* dq *passing through a cross-sectional area* A *divided by the time* dt *it takes for the charge to pass through the area.*

$$I \equiv \frac{dq}{dt}$$

The sign ≡ means "per definition equal to".

The total amount of charge q that has passed through the cross section is a function of time $q = q(t)$. If the charge q is given as a function of time, $q = q(t)$, we may differentiate $q(t)$ with respect to time in order to find the current.

$$I \equiv \frac{dq}{dt} = q'(t) \tag{3.2}$$

The charge is quantified

$$q = z \cdot n \cdot e\ ,\ where\ n \in N_o,\quad i.e.\ n = 0, 1, 2, 3, ... \tag{3.3}$$

Here,

- z is the integer *charge number*, e.g., $z = 2$ for Ca^{2+} and $z = -1$ for an electron e^-.
- n is the number of charged particles under consideration.
- e is one *quantum of charge*, the smallest possible amount of charge that we can have,

$$e = 1,602 \cdot 10^{-19}\ \mathrm{C}$$

Here C is the unit of charge (coulomb). The unit of current is Ampere = coulomb per second or $A = \frac{C}{s}$.

The Avogadro constant is equal to the number of atoms per mole.

$$N_A = 6,022 \cdot 10^{23}\ \mathrm{mole}^{-1} \tag{3.4}$$

One mole of charge quanta therefore corresponds to a total charge which is called *Faraday's constant*.

$$1\ F = N_A \cdot e = 96\,485\ \mathrm{C/mole} \tag{3.5}$$

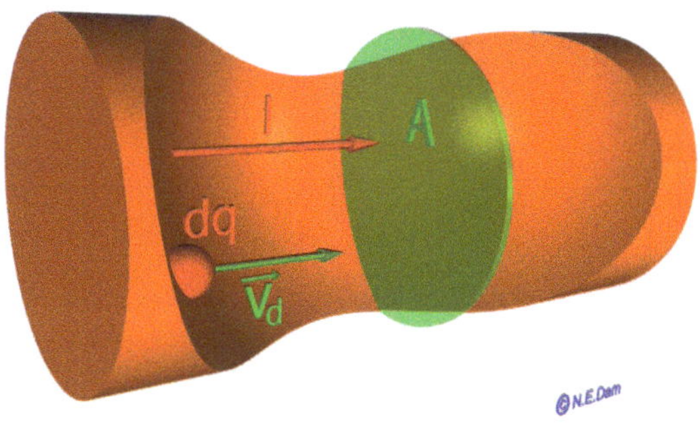

Fig. 3.1 Definition of the electric current through a cross-sectional area A of a wire. An amount of charge dq passes through an area A during the time dt. The drift velocity is denoted $\mathbf{v_d}$

The current reference direction

The positive current direction is indicated using an arrow as shown in Fig. 3.1.

- *The current is positive* if the positive charge $dq > 0$ flows in the positive current direction indicated by the arrow or if the negative charge flows in the opposite direction.
- *The current is negative* if the negative charge $dq < 0$ flows in the positive current direction indicated by the arrow or if the positive charge flows in the opposite direction.

When we analyse a circuit, we usually don't know the current directions to begin with. However, we chose some positive reference directions indicated by arrows, as shown in Fig. 3.2. The directions are chosen arbitrarily. When we have chosen the directions, we should not change them. In addition to the directions, we assign current variables I_1, I_2, I_3, ... When we calculate the circuit, we find some values for these currents. If I_1 is positive, the positive charge flow is in the same direction as indicated by the arrow next to I_1. If I_2 is negative, the positive charge flow is in the opposite direction of the arrow next to I_2.

An alternative way to indicate the current reference direction:
I_{ab} is the current from a to b. The reference direction (the arrow) is from a to b.
If I_{ab} is positive, the positive charge flows from a to b.

In normal metals and superconductors, only negative electrons can transport charge through the crystal structure. The positive charge is attached to the atomic nuclei, which cannot move because they are fixed in the crystal lattice.

For example, if we consider Fig. 3.2 saying that "the value for I_1 is positive and the positive charge therefore flows in the same direction as indicated by the arrow above I_1", then it is the negatively charged electrons that flow in the opposite direction of the arrow.

In liquid electrolytes and solid-state ion conductors, both positive and negative ions contribute to charge transport.

Charge accumulation in a node is not possible

For example, if we consider Fig. 3.2 again and consider node a, we have the following:

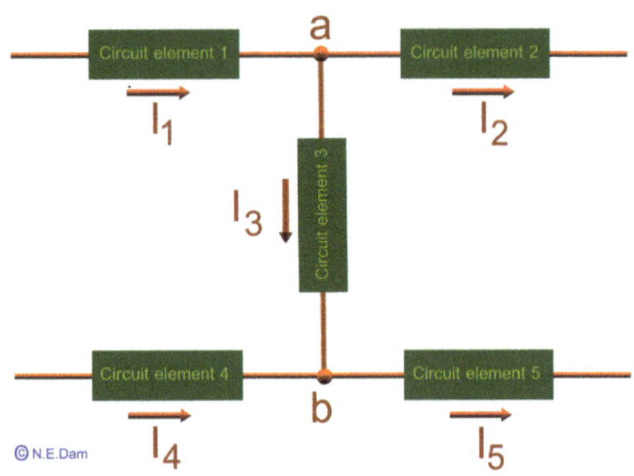

Fig. 3.2 Current directions in a part of a circuit

According to Eq. 3.2, the current I_1 during time dt will deliver the charge $I_1\ dt$ to the node a. During the same time period dt, the currents I_2 and I_3 carry the charges $I_2\ dt$ and $I_3\ dt$ away from node a. Since the charge cannot accumulate at the node a, the same amount of charge must be removed as is added:

$$\begin{aligned} I_1\ dt = I_2\ dt + I_3\ dt \Rightarrow \\ I_1 = I_2 + I_3 \end{aligned} \tag{3.6}$$

This is generalised in

Theorem 2 ***Kirchhoff's Current Law (KCL).*** *The sum of currents in a node is equal to the sum of currents away from the same node.*

$$\Sigma_i I_i \mathit{in\ a\ node} = \Sigma_j I_j \mathit{away\ from\ the\ same\ node}$$

This theorem does not apply if there is a capacitor connected to the node under consideration, as the charge can then accumulate in the capacitor.[5]
In a later section, we will consider how to apply Kirchhoff's Current Law.

AC/DC current

- *DC means Direct Current.* It is when the current flows in the same direction all the time. Quite often, the current has a constant value.

[5] In the extended version of Ampere's law in electrophysics, an electrical flux current (formerly called displacement current) is introduced. If this flux current is included, KCL can also be used if there is a capacitor connected to the node under consideration.

- *AC means Alternating Current.* It is when the current alternates between flowing in one direction and in the opposite direction. Usually, the current is a periodic function of time.

Current density

The current density is defined as the current I per area A, see Fig. 3.1

$$J \equiv \frac{I}{A} \tag{3.7}$$

The area A should be perpendicular to the current flow direction.

3.2 Voltage

The voltage is a measure of the difference in the electrochemical potential of electrons.

A voltmeter measures this difference between two points a and b. *The electrochemical potential is the sum of a chemical potential (for an electron) and a constant times an electrostatic potential (also for an electron).* To simplify the introductory description in this section we will ignore the influence of chemical potentials, i.e. we assume that we have the same chemical potential everywhere. Thus, in this section, the voltage is reduced to the *electrostatic voltage equal to the difference in the electrostatic potential.*

To understand the concept of electrostatic voltage, we must first consider the concepts of electrical potential energy and then electrical potential.

3.2.1 Electrical Potential Energy

We consider two charges q_o and q_1 with the distance between them r_{o1}. In Electrophysics it is shown that the two charges achieve an electric potential energy given by

$$U_{epot} = \frac{q_o q_1}{4\pi \epsilon_o r_{o1}} \tag{3.8}$$

Here, the constant ϵ_o is the vacuum permittivity

$$\epsilon_o = 8,854 \cdot 10^{-12} \frac{\mathrm{As}}{\mathrm{Vm}} \tag{3.9}$$

The zero-point of electric potential energy is when the distance between the two charges approaches infinity. As the distance r_{o1} between the two charges approaches infinity, the electric potential energy approaches zero, cf. Equation 3.8.

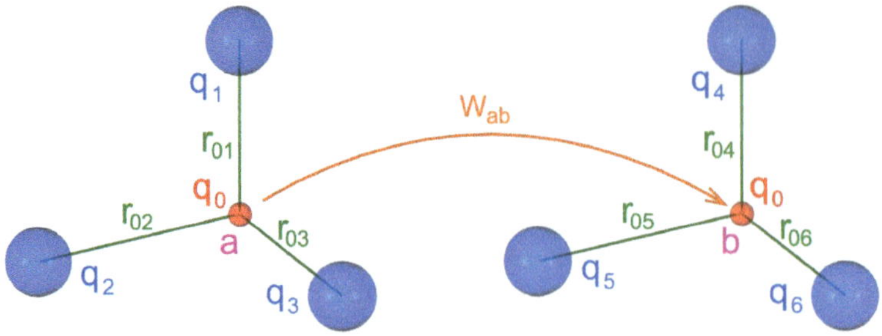

Fig. 3.3 The figure is used when calculating the electric potential energy of a charge q_o relative to the three charges q_1, q_2, q_3, when the charge q_o is located at the point a. It is assumed that q_o is so far away from the charges q_4, q_5, q_6 that we may neglect the electric potential energy that q_o obtain relative to q_4, q_5, q_6. N.E.Dam ©

Figure 3.3 shows an electric charge q_o at the point a. The charge q_o is surrounded by three charges q_1, q_2, q_3. As a result, q_o acquires electric potential energy relative to each of the three charges

$$U_{epot,a} = \frac{q_o q_1}{4\pi\epsilon_o r_{o1}} + \frac{q_o q_2}{4\pi\epsilon_o r_{o2}} + \frac{q_o q_3}{4\pi\epsilon_o r_{o3}} = \frac{q_o}{4\pi\epsilon_o}\{\frac{q_1}{r_{o1}} + \frac{q_2}{r_{o2}} + \frac{q_3}{r_{o3}}\}$$
$$= \frac{q_o}{4\pi\epsilon_o}\sum_{i=1}^{i=3}\frac{q_i}{r_{oi}} \tag{3.10}$$

This is the electric potential energy of q_o relative to the three charges q_1, q_2, q_3. The expression can, of course, be extended to apply if there are more than three charges in the environment of q_o. If there are n charges in the environment of q_o, it becomes

$$U_{epot,a} = \frac{q_o}{4\pi\epsilon_o}\sum_{i=1}^{i=n}\frac{q_i}{r_{oi}} \tag{3.11}$$

When we build the system of the three charges q_1, q_2, q_3 (without q_o), we bring the three charges q_1, q_2, q_3 from infinitely far apart to the positions shown in Fig. 3.3. In this way, the three charges q_1, q_2, q_3 achieve an electric potential energy relative to each other without q_o being present. In this context, we are not interested in this electric potential energy. We are only interested in what happens when q_o is brought from infinitely far away and to position a between the three charges q_1, q_2, q_3. Then q_o achieves an electric potential energy relative to the three charges q_1, q_2, q_3. This electric potential energy is the one we have calculated in Eq. 3.10.

The charge q_o can be located at point a or point b.

When q_o is located at point a, the electric potential energy is calculated for q_o relative to q_1, q_2, q_3. The distances from q_o to q_4, q_5, q_6 are assumed to be so large (relative to

the distances from q_o to q_1, q_2, q_3) that it is a good approximation to ignore the electric potential energy between q_o and q_4, q_5, q_6.
When q_o is located at point b, the electric potential energy is calculated for q_o relative to q_4, q_5, q_6, neglecting the electric potential energy for q_o relative to q_1, q_2, q_3.

$$U_{epot,b} = \frac{q_o}{4\pi\epsilon_o}\sum_{i=4}^{i=6}\frac{q_i}{r_{oi}} \tag{3.12}$$

It requires work W_{ab}, to move q_o from position a to position b. Without going deeper into electrophysics, it can be stated that the work done by an electric force is equal to the decrease in the electric potential energy

$$W_{ab} = U_{epot,a} - U_{epot,b} \tag{3.13}$$

The left side of this equation represents work, such as that done by an electric force acting on q_o while moving it from point a to point b. At the same time, there is a decrease in electric potential energy corresponding to the right side of this equation.

An analogy might be a cart on a roller coaster. In this case W_{ab} is the work that the gravitational force does on the cart when it is pulled down from a high point a to a low point b. At the same time, there is a decrease in potential energy

$$W_{ab} = U_{pot,a} - U_{pot,b} \tag{3.14}$$

3.2.2 Electrical Potential

In the above section, we saw how to calculate the electric potential energy $U_{epot,a}$ for a charge q_o located at the point a near the charges q_1, q_2, q_3. We assumed that the distance from q_o to q_4, q_5, q_6 was so great that we could ignore the electric potential energy of q_o relative to q_4, q_5, q_6.

If we assume that q_o is at the point a, then $U_{epot,a}$ is the electric potential energy of q_o relative to q_1, q_2, q_3. If we divide $U_{epot,a}$ by q_o we get the *electric potential* at the point a

$$V_a \equiv \frac{U_{epot,a}}{q_o} = \frac{1}{4\pi\epsilon_o}\sum_{i=1}^{i=3}\frac{q_i}{r_{oi}} \tag{3.15}$$

The electric potential is the electric potential energy of $\mathrm{q_o}$ *(relative to surrounding charges) divided by* $\mathrm{q_o}$. *Note that the potential does not depend on q_o as we have divided by* $\mathrm{q_o}$. *The potential depends on where* $\mathrm{q_o}$ *is located, i.e., it depends on where point* a *is located.*

If we change the location of the point a, the distances from q_o to q_1, q_2, q_3 change. This changes both the electrical potential energy and the potential, cf. Eqs. 3.10 and 3.15.

At point b, we find the potential

$$V_b \equiv \frac{U_{epot,b}}{q_o} = \frac{1}{4\pi\epsilon_o}\sum_{i=4}^{i=6}\frac{q_i}{r_{oi}} \tag{3.16}$$

3.2.3 Definition of Voltage

If we divide Eq. 3.13 by q_o, we find the following

$$\frac{W_{ab}}{q_o} = \frac{U_{epot,a}}{q_o} - \frac{U_{epot,b}}{q_o} = V_a - V_b \equiv V_{ab} = V \tag{3.17}$$

Here, $V_a - V_b$ is the drop in potential from point a to point b. This drop in potential is denoted V_{ab}. In this context, the voltage V is equal to the potential drop V_{ab}. The potential drop V_{ab} can be positive or negative:

$$V_a > V_b \Rightarrow V = V_{ab} = V_a - V_b > 0 \text{ (a drop in potential from a to b)} \tag{3.18}$$

$$V_a < V_b \Rightarrow V = V_{ab} = V_a - V_b < 0 \text{ (an increase in potential from a to b)} \tag{3.19}$$

If q_o is an infinitesimal small charge dq, the work becomes dW_{ab} and we arrive at the general definition of voltage

Theorem 3 ***The definition of voltage*** *is the work* $\mathrm{dW_{ab}}$ *that must be performed to move the charge* dq *from* a *to* b *divided by* dq*:*

$$V_{ab} \equiv \frac{dW_{ab}}{dq} \text{ or for short } V \equiv \frac{dW}{dq} = -\frac{dU}{dq}$$

The voltage is equal to the difference in potentials at points a *and* b*:*

$$V = V_{ab} = V_a - V_b = \frac{U_a}{dq} - \frac{U_b}{dq}$$

Remarks on the following relationship

$$V \equiv \frac{dW}{dq} = -\frac{dU}{dq} \tag{3.20}$$

This relationship is due to the fact that the work done (dW) is equal to the decrease in potential energy, which is equal to minus the increase in potential energy ($-dU$). This applies in general, i.e., not only for an electric force and an electric potential energy as in Eq. 3.13, but also where the force is a result of both electric forces and chemical processes.

In that case, the potential energy U will include both the electric potential energy and the chemical potential energy.

Theorem 3 is therefore also valid when V_a and V_b are electrochemical potentials. In this case, U_a and U_b are the total potential energies that include both chemical and electrical potential energies at the points a and b. If the chemical potential is constant, then the voltages V and V_{ab} become a difference between the electrical potentials.

3.2.4 Definition of the Electron Volt

If an elementary charge (one *charge quantum* e) is accelerated from a to b by an electric force through a potential drop of one volt, then according to Theorem 3 the electric force does the following work (J = joule, C = coulomb and V = volt $= \frac{\text{joule}}{\text{coulomb}}$)

$$\begin{aligned} dW_{ab} &= V_{ab}\,dq = 1\text{ V} \cdot 1\,e = 1\text{ eV (one electron volt)} \\ 1\,e\text{V} &= 1\text{ V} \cdot 1\,e = 1\,\frac{\text{J}}{\text{C}} \cdot 1,602 \cdot 10^{-19}\text{C} = 1,602 \cdot 10^{-19}\text{J} \end{aligned} \tag{3.21}$$

i.e., the work done by the electric force is equal to the voltage V_{ab} times the charge dq, that is, 1 volt times the charge quantum e. This is called one electron volt. As also shown in Eq. 3.20, then

$$dW = -dU \tag{3.22}$$

i.e. the work is equal to the drop in potential energy. This drop in potential energy is accompanied by an equally increase in the kinetic energy for the elementary charge considered. We assume that the particle does not lose kinetic energy by colliding with other particles.

Theorem 4 ***The definition of one electron volt (1 eV)*** *is the drop in potential energy when an elementary charge (one charge quantum e) is accelerated from a to b by an electric force through a potential drop of one volt*

$$dW_{ab} = -dU = V_{ab}\,dq = 1\text{ V} \cdot 1\,e = 1\text{ eV}$$

When working with energy levels for atoms and molecules, it is more convenient to use electron volt as a unit rather than joule. For example, it is easier to say that "the electron energy level increases by 2 eV" than to say "the electron energy level increases by $3,2 \cdot 10^{-19}$J".

3.2.5 Definition of the Electrochemical Potential

In this section, we define the electrochemical potential. It is a little more advanced than the rest of the book and can be skipped on the first reading because one can easily calculate many common electrical circuits without knowing the electrochemical potential. However, it is absolutely necessary to know the electrochemical potential if one wants to understand the following:

- Why do the diode and other semiconductor components work as they do?
- Why can a potential difference arise between two different metals?
- How electrochemical cells such as batteries work

Theorem 5 ***The electrochemical potential*** *is defined as the sum of two terms, where the first term is the chemical potential, and the second term includes the electrostatic potential.*

$$\tilde{\mu}_i = \mu_i + z_i \; F \; \Phi \qquad \text{unit : joule/mole} \tag{3.23}$$

This definition includes the following quantities[6]*:*

- $\tilde{\mu}_i$ *is the electrochemical potential of the substance* i.
  The electrochemical potential has a term with the chemical potential and a term with the electrical potential. The unit is joule per mole (J/mol).
- μ_i *is the chemical potential of the substance* i.
  The chemical potential of substance i *depends on the chemical environment in which substance* i *is found. More precisely, we have the following definition of the chemical potential of substance* i*:*
  We consider a system with a very large amount of a mixture of several substances $j = 1, 2, 3, \ldots$*. We keep the temperature and pressure constant in the system and in equilibrium in temperature and pressure with the environment. Furthermore, we keep the amounts of all substances (except substance* i*) constant. The total chemical potential energy of the system in this case of constant temperature and pressure is expressed by G (Gibbs free energy). Since the amount of the system is assumed to be very large, the mixing ratio of the substances does not change even if we add 1 mole of substance* i*. The chemical potential of substance* i *is then defined as the increase in* G *when we add 1 mole of substance* i *to a very large amount of a mixture, and the unit is joule per mole* (J/mol).
- z_i *is the charge number of the substance* i.
  Examples of substances can be Na^+, Cl^-, Zn^{2+}, e^- *that have charge numbers* $z_i = 1, -1, 2, -1$.
- *Faraday's constant is the amount of charge for one mole of charge quanta, as defined in Eq. 3.5 on page 54.*
- Φ *is the electrostatic potential with unit volt. In other sections of this book, where we disregard the chemical potential, we use, for example, the designation* V_a *for the elec-*

[6] It is beyond the scope of this book to define these quantities in detail, it would require many pages of text and figures, and it belongs in electrophysics and chemistry.

trostatic potential at the point a. However, in this section, where we consider both the chemical potential and the electrostatic potential, the symbol Φ_a is used for the electrostatic potential at the point a.

Now, we consider the second term of the electrochemical potential, i.e., the term with the electrostatic potential Φ. When we multiply Φ by Faraday's constant, which is the number of coulombs per mole, we get the unit

$$\frac{\text{joule}}{\text{coulomb}} \frac{\text{coulomb}}{\text{mole}} = \frac{\text{joule}}{\text{mole}}$$

While the unit for Φ is volt, the unit for $F\ \Phi$ is joule per mole (J/mol), i.e., the same unit as for the chemical potential μ_i .

If we divide Eq. 3.23 by F, then all terms in the equation have the unit volt.

$$\frac{\tilde{\mu_i}}{F} = \frac{\mu_i}{F} + z_i \Phi \qquad \text{unit : volt} \tag{3.24}$$

If we multiply Eq. 3.24 by one charge quantum e, then all terms in the equation have the unit electron volt.

$$e\frac{\tilde{\mu_i}}{F} = e\frac{\mu_i}{F} + z_i e \Phi \qquad \text{unit : electron volt} \tag{3.25}$$

If we consider the electrochemical potential of a positive charge quantum e^+, then Eq. 3.24 becomes

$$\frac{\tilde{\mu}_{e^+}}{F} = \frac{\mu_{e^+}}{F} + \Phi \qquad \text{unit : volt} \tag{3.26}$$

If we consider the electrochemical potential of a negative charge quantum e^-, then Eq. 3.24 becomes

$$\frac{\tilde{\mu}_{e^-}}{F} = \frac{\mu_{e^-}}{F} - \Phi \qquad \text{unit : volt} \tag{3.27}$$

The term $e\ \mu_{e^-}/F$ is, to a good approximation, the Fermi energy[7] with unit electron volt (eV). It is the maximum kinetic energy of the conduction electrons in a "potential well model" when the bottom of the well is taken to be at zero kinetic energy. According to reference[8] it is about 7 eV for copper. The term $e\ \tilde{\mu}_{e^-}/F$ is the electrochemical potential energy with unit electron volt (eV). It corresponds to the Fermi level, which is the Fermi energy superimposed on the electrostatic potential energy.

Example: The change in electrochemical potential across a resistor

We consider a copper wire with some resistance from a to b, see Fig. 3.4. The copper wire is connected to a 1,5 V electrochemical cell (also called somewhat incorrectly a battery). We neglect the resistance in the connecting wires, which are also made of copper.

[7] Strictly speaking, Fermi energy is defined at absolute zero temperature $T = 0\ K$.

[8] Philos Trans A Math Phys Eng Sci (1957) 250 (979): 325âŁ"357.

We consider the following graphic representation:

1. The top horizontal line is the chemical potential (measured in volt) of copper. It is, of course, constant throughout the material.
2. The line with a negative slope is the electrostatic potential Φ. It is 0 V at the point b, which is connected to the ground and, by definition, has zero electrostatic potential. At point a, the electrostatic potential is $\Phi = 1,5$ V.
3. According to Eq. 3.27, which is the electrochemical potential of an electron, the electrostatic potential Φ must be subtracted from the chemical potential μ_{e^-}/F. This gives the line with positive slope, i.e., the electrochemical potential $\tilde{\mu}_{e^-}/F$. This line shows that the electron can reduce its electrochemical energy by moving from b to a. This is what happens physically with the electrons in both the resistor and the wires to and from the electrochemical cell (Fig. 3.5).

We continue with Fig. 3.4 which is physically correct. When we measure with a voltmeter, it is in contact with the conduction electrons. Since the driving force is towards the lower electrochemical potential of the conduction electron, it moves to the left from b to a as shown in Fig. 3.4. When one mole of electrons moves from b to a, the drop in the electrochemical potential energy per mole is $\tilde{\mu}_{e^-,b} - \tilde{\mu}_{e^-,a}$. Since voltage is defined as the drop in potential energy per charge, we find

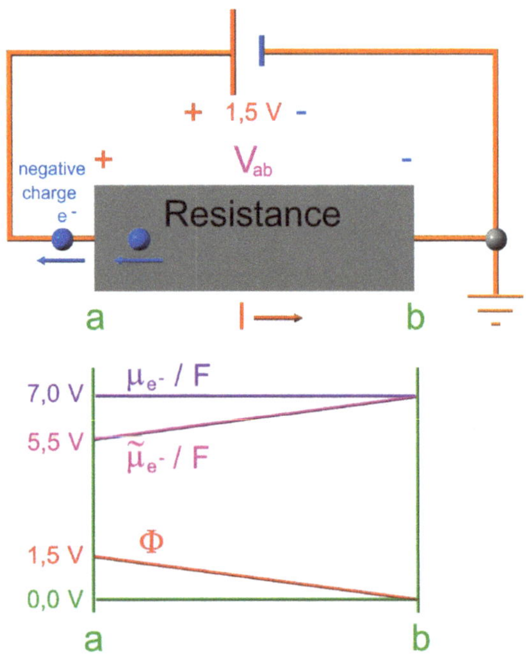

Fig. 3.4 The change in electrochemical potential for a negative charge quantum across a resistor. N.E.Dam ©

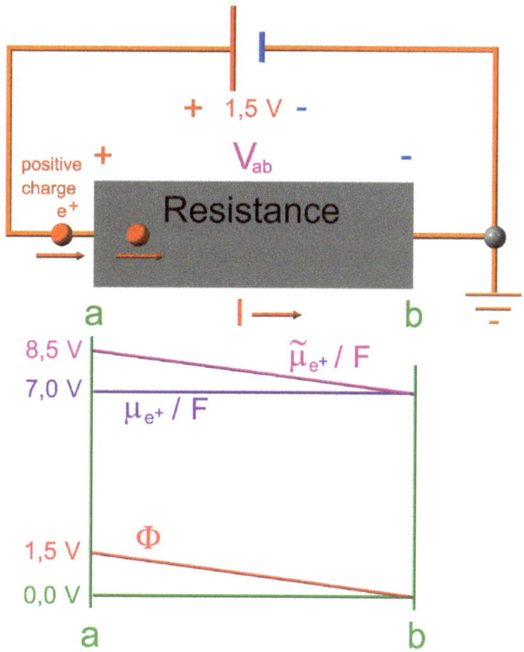

Fig. 3.5 The change in electrochemical potential for a positive charge quantum across a resistor. N.E.Dam ©

$$V_{ab} = \frac{\tilde{\mu}_{e^-,b} - \tilde{\mu}_{e^-,a}}{F} \tag{3.28}$$

Substituting Eq. 3.27 into Eq. 3.28 we find the following equation for the voltage

$$V_{ab} \equiv V_a - V_b = \left(\frac{\mu_{e^-,b}}{F} - \Phi_b\right) - \left(\frac{\mu_{e^-,a}}{F} - \Phi_a\right) \tag{3.29}$$

This expression is valid for electrochemical cells where the chemical potential for electrons is different at the two electrodes, i.e., $\mu_{e^-,a} \neq \mu_{e^-,b}$.

With constant chemical potential, as is the case if we have the same material all the way from a and b, then $\mu_{e^-,a} = \mu_{e^-,b}$ and 3.29 becomes

$$V_{ab} = V_a - V_b = \Phi_a - \Phi_b \tag{3.30}$$

and

$$V_a = \Phi_a \text{ and } V_b = \Phi_b \tag{3.31}$$

We now consider Fig. 3.4.

As mentioned earlier, there are no positive charges moving in a normal metal such as copper. However, when negative electrons move to the left from b to a, this corresponds to fictitious positive charges moving to the right from a to b. If we describe the current using fictitious positive elementary charges, we must use Eq. 3.26. This is depicted in Fig. 3.4. The positive charge quantum moves to the right from a to b and the electrochemical potential drops $1,5$ V from $8,5$ V to $7,0$ V.

Although it is not physically correct, it is a possible model to consider positive charges moving through the material in the resistor. In order for this to be realised, we may consider a P-type semiconductor where charge transport is largely carried out by positive holes. Here, the Fermi energy will have a different (lower) value than for copper, i.e. the horizontal line at $7,0$ V in Fig. 3.4 is at a lower level, but still horizontal. The line from $8,5$ V to $7,0$ V will move correspondingly downward, so there will still be the same drop in the electrochemical potential.

If we consider Fig. 3.4 a positive charge e^+ moves from a to b through a drop in the electrochemical potential, which is $\tilde{\mu}_{e^+,a} - \tilde{\mu}_{e^+,b}$. Since voltage is defined as the drop in potential energy per charge, we find

$$V_{ab} = V_a - V_b = \frac{\tilde{\mu}_{e^+,a} - \tilde{\mu}_{e^+,b}}{F} \tag{3.32}$$

If $\mu_{e^+,a} = \mu_{e^+,b}$ we find the following in the same way as described above for electrons

$$V_{ab} = V_a - V_b = \Phi_a - \Phi_b \tag{3.33}$$

and

$$V_a = \Phi_a \text{ and } V_b = \Phi_b \tag{3.34}$$

Even though the model with positive charges does not match physical reality, we end up with the same result as with electrons as charge carriers.

3.2.6 Current Flow Through a Resistor

We consider the circuit element in Fig. 3.6. In this case, it is a resistor with resistance R. The current I_{ab}, which flows through the circuit element from a to b, is considered to be positive. Therefore, a positive charge dq flows through the circuit element in the direction indicated by the arrow.[9] The potential[10] at point a is V_a. The potential at point b is V_b. If a

[9] As mentioned in the previous section, positive charge does not flow through a normal metal, if the current is positive, we pretend that it is positive charge flowing in the direction of the current arrow, knowing that it is negatively charged electrons that flow in the opposite direction of the arrow.

[10] When we use the words potential and potential energy in this section, where we disregard changes in chemical potentials, it can be understood as electrostatic potential and electrical potential energy, respectively. More generally, if we do not disregard changes in chemical potentials, then the word potential represents electrochemical potential.
When we use the word potential at point a, we mean the difference in potential between point a and a chosen zero point.

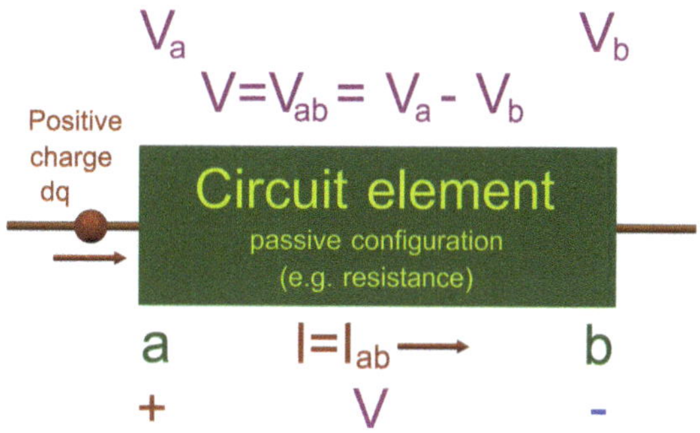

Fig. 3.6 The potential drop V_{ab} when the current I_{ab} flows through a circuit element. N.E.Dam ©

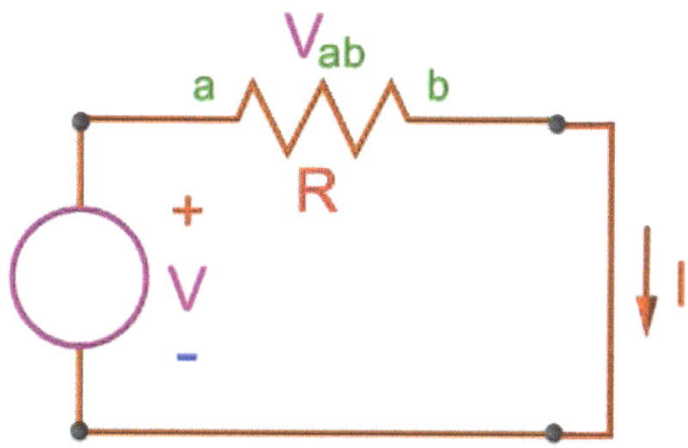

Fig. 3.7 A simple circuit with a voltage generator and a resistor. N.E.Dam ©

positive charge dq flows through the resistor from a to b, the drop in potential V_{ab} will be equal to the drop in energy[11] from a to b divided by charge dq

$$V_{ab} \equiv \frac{U_a - U_b}{dq} \tag{3.35}$$

This equation is analogous to Eq. 3.17.

Figure 3.7 shows a simple closed circuit with a voltage generator V and a resistor R. When a positive charge q_o passes through the voltage generator from - to +, its potential increases by V. When q_o passes through the resistor R, the potential of q_o decreases by $V = V_{ab}$. The force pulling q_o through the resistor does work W_{ab}, which is equal to the loss of energy. Thereby we find

[11] When we use the word potential energy, or just energy, in this section, where we disregard changes in chemical potentials, it can be understood as electrical potential energy. More generally, if we do not disregard changes in chemical potentials, then the word "energy" is the total energy U, which with a very good approximation is the sum of electrical potential energy and chemical potential energy.

$$V_{ab} \equiv \frac{U_a - U_b}{q_o} = \frac{W_{ab}}{q_o} \tag{3.36}$$

3.2.7 Changes in Potential Around a Simple Electric Circuit

Figure 3.8 shows the *potential in a circuit* with a battery, V_{Bat}, an internal resistance R_i and an external resistor R. We define the points a and d as having zero potential. When we move around with positive charge q_o in the closed circuit, we observe the following variations of the potential V:

- Moving through the electrochemical cell from a to b, the potential V increases from 0 to V_{Bat}. The electrochemical reactions 'force' the positive charge q_o to be pushed towards the plus pole of the battery (which corresponds to the situation in which we push a ball uphill). Thereby, the energy increases (as the gravitational potential energy increases for the ball on a hill). Hence, the potential V (which is the energy U divided by q_o) also increases. The increase in energy U is denoted $\Delta_a^b U \equiv U_b - U_a$.
- When q_o exits the electrochemical cell at point b, it has higher energy than at point a and, therefore, also higher potential V. However, some of its energy is lost in the internal resistance R_i from point b to point c. In fact, this happens already when q_o passes through the electrolyte in the electrochemical cell (from a to b). The electrolyte is the substance with ions between the negative and positive pole. At the same time that q_o increases its energy, it also loses a minor fraction of the energy when q_o bumps into ions in the electrolyte. The result is that when q_o exits the battery at point c, it has increased its potential by V_{Bat} and, at the same time, decreased its potential by $R_i I$, where I is the current flow through the battery. The increase in energy U through internal resistance R_i becomes $\Delta_b^c U \equiv U_c - U_b$. This increase is negative, as it is actually a decrease.
- It is not physically possible to divide into $a - b$ and $b - c$, since the increase in potential and the decrease in potential due to R_i occur simultaneously at the same place in the electrochemical cell. The division into $a - b$ and $b - c$ is an equivalent electrical diagram that should visualise what happens in the battery.
- Outside the battery, the potential drop from c to d is RI. Subsequently, the potential level at d is zero, $V_d = 0$. The increase in energy U through external resistance R becomes $\Delta_c^d U \equiv U_d - U_c$. This increase is negative because it is actually a decrease in energy.

If we add all changes in energy, the sum will be equal to zero as we return to the same zero point $d = a$. We find

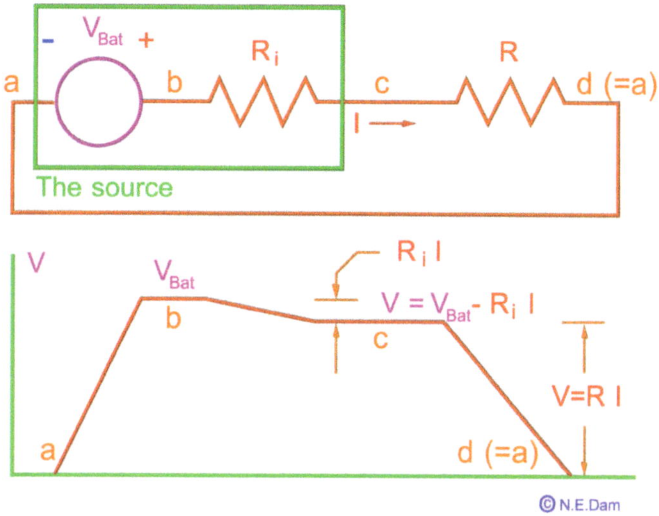

Fig. 3.8 The changes of potential around a simple electric circuit with a battery. N.E.Dam ©

$$\begin{aligned} \Delta_a^b U + \Delta_b^c U + \Delta_c^d U = 0 \Rightarrow \\ (U_b - U_a) + (U_c - U_b) + (U_d - U_c) = 0 \Rightarrow \\ \frac{U_b - U_a}{q_o} + \frac{U_c - U_b}{q_o} + \frac{U_d - U_c}{q_o} = 0 \Rightarrow \\ V_{ba} + V_{cb} + V_{dc} = 0 \end{aligned} \tag{3.37}$$

The first term is positive, since it is actually an increase in energy. The second and third terms are both negative since it is actually a decrease in energy. If we move the two negative terms to the right side of the equal sign, a minus sign appears in front of them. The minus sign disappears if we change the order cb to bc and dc to cd.

$$\begin{aligned} V_{ba} = -V_{cb} - V_{dc} \Rightarrow \\ V_{ba} = V_{bc} + V_{cd} \end{aligned} \tag{3.38}$$

On the left side, we have the potential increase in the circuit. On the right side, we have the sum of the potential decreases in the same circuit. This leads to Kirchhoff's voltage Law (KVL):

Theorem 6 ***Kirchhoff's voltage Law (KVL)*** *states that if we move around a closed electrical loop, the sum of potential increases will equal the sum of potential decreases.*

$$\Sigma_i V_i \text{ (potential increases in a loop)} = \Sigma_j V_j \text{ (potential decreases in a loop)}$$

In a later section, we will consider how to apply Kirchhoff's voltage Law.

3.3 Power

If we divide the work dW by the time interval dt in which the work is performed, we get the rate at which the work is performed, i.e., the work per time. This is precisely the *definition of power*.

Theorem 7 ***The definition of power is*** *the work performed per time*

$$P \equiv \frac{dW}{dt} = -\frac{dU}{dt}$$

Power can also be considered as a change in energy per time, i.e., the rate at which energy is converted.

If we multiply this fraction by $\frac{dq}{dq}$ we find the following

$$P = \frac{dW}{dt}\frac{dq}{dq} = \frac{dW}{dq} \cdot \frac{dq}{dt} \tag{3.39}$$

We have already defined the current as follows, cf. Theorem 1 on page 54

$$I \equiv \frac{dq}{dt} \tag{3.40}$$

We have also defined the voltage as follows, cf. Theorem 3 on page 60

$$V \equiv \frac{dW}{dq} \tag{3.41}$$

If we substitute Eqs. 3.40 and 3.41 into Eq. 3.39, we find the following

$$P \equiv V \cdot I \tag{3.42}$$

In this equation, the voltage V is the drop in potential, which is the drop in energy per charge (joule per coulomb). Current I is the amount of charge that passes through the circuit element per second (coulomb per second). The SI unit for power becomes joules per second, also denoted watt or W = J/s. The power can vary over time.

3.4 Resistance and Conductance

Throughout this section, we assume that the temperature is kept constant.

Theorem 8 ***Definition of Ohm's law.*** *The voltage* V *across a resistance* R *is equal to the product of the resistance* R *and the current* I *through the resistance*

$$V = R \cdot I \tag{3.43}$$

In this case, we use the sign convention for current and voltage shown in Fig. 3.6, which shows the passive configuration. In this figure, the current is considered positive when it flows through the resistor in the same direction as the potential drops.

NB! Eq. 3.43 denotes the Ohms law if and only if R is independent of the current I! If R depends on the current I, then Eq. 3.43 is not denoted by Ohm's law. However, Eq. 3.43 can still be used.

Equation 3.43 can be used no matter whether R is constant or not.

- If R is constant and independent of current I, Ohm's law is valid, and we consider an Ohmic material. In this case, the voltage V is a linear function of the current I. The straight red line $V = RI$ through $(0, 0)$ in Fig. 3.9 is an Ohmic material because the slope R is constant and independent of I. Quite often, Ohmic materials are used as resistors.
- If R depends on the current I, Ohm's law is not valid and we consider a non-Ohmic material. However, Eq. 3.43 can still be used as a current-voltage relationship. Many materials have resistance R, which depends on current I. Ohm's law is not valid for such non-Ohmic materials. However, the relationship between voltage V and current I in Eq. 3.43 is still valid. Perhaps it seems strange that this relation may also be used in the case where Ohm's law is not valid. The reason is that Ohm's law is a question of whether R is constant or not.

For the non-Ohmic material in Fig. 3.9 we consider the point P_1 with the coordinates (I_1, V_1). If we take the ratio between V_1 and I_1 at point P_1 it is possible to define the *static resistance*, which is the static slope of the straight line from (0,0) to P_1.

At point P_1, we find, for example, that the static resistance is determined by

$$R_{S1} = \left(\frac{V}{I}\right)_{I=I_1} = \frac{V_1}{I_1} \tag{3.44}$$

If we consider another point P_2, we may find the static slope $R_{S2} = V_2/I_2$, which is the static resistance at the point P_2. Apparently, the static slope (the static resistance) increases from P_1 to P_2. Ohm's law is not valid for this material, but the relation in Eq. 3.43 is still valid and can be used to find the static resistance at any point in the V(I) curve for the non-Ohmic material.

It is also possible to define the *dynamic resistance*, which is the tangent slope at a given point on the curve $V(I)$.

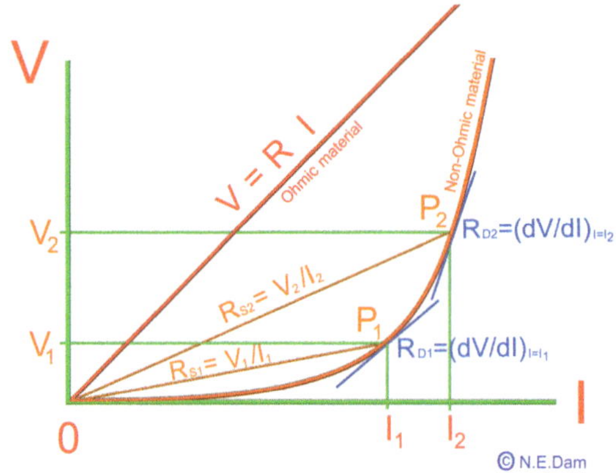

Fig. 3.9 The voltage V as a function of the current I for an Ohmic material and a non-Ohmic material. R_{S1} and R_{S2} are the static resistances at points P_1 and P_2. R_{D1} and R_{D2} are the dynamic resistances at points P_1 and P_2

At point P_1, we find, for example, that the dynamic resistance is determined by

$$R_{D1} = \left(\frac{dV}{dI}\right)_{I=I_1} \tag{3.45}$$

We will later encounter the dynamic resistance in connection with the current-voltage characteristic of a diode.

Resistance of a wire

The resistance of a wire is qualitatively speaking the wire's ability to resist a current flow. A high value of the resistance means that it is 'difficult' for the current to flow through the wire. Quantitatively speaking, the resistance R_{ab} of a wire from a to b is the voltage V_{ab} along the wire from a to b divided by the current I_{ab} flowing from a to b. This is implied in Eq. 3.43 or it can be expressed as

$$R_{ab} = \frac{V_{ab}}{I_{ab}} \tag{3.46}$$

If we cut or remove the wire, we get an infinitely high resistance, $R_{ab} \to \infty$. If we replace the wire by a superconducting wire, we get a zero value for the resistance, $R_{ab} = 0$. In the circuit diagrams, we neglect the resistances of the wires we draw between the circuit elements. This means that $R_{ab} = 0$ on all the wires between the circuit elements.

The resistance of a normal wire (not a superconducting wire) is proportional to the length L of the wire and inversely proportional to the cross-sectional area A of the wire. This may be expressed in the following equation

$$R = \rho \, \frac{L}{A} \tag{3.47}$$

The proportionality constant ρ is the *resistivity*. For copper, the resistivity is $\rho = 1,72 \cdot 10^{-8}$ Ωm. The Greek letter 'rho' can be written as ρ or as ϱ.

The resistivity is independent of the length and independent of the cross-sectional area of the wire. The resistivity depends only on the material and the temperature.

If we compare the resistance of different wires made of different materials, the difference in resistance may be due to different lengths or different cross-sectional areas. Therefore, when we compare different materials, we should compare their resistivities, because they are independent of the lengths and cross-sectional areas of the wires.

If we substitute $L = 1\,\text{m}$ and $A = 1\ \text{m}^2$ in Eq. 3.47, we find that resistance $R = \rho\,\frac{1\,\text{m}}{1\,\text{m}^2}$ means that the resistivity is numerically the resistance of a wire, which is 1 m long and 1 m^2 in the cross-sectional area.

Conductance of a wire

The conductance of a wire is, qualitatively speaking, the wire's ability to conduct a current flow. A high value of conductance means that it is easy for current to flow through the wire. Quantitatively speaking, the conductance G_{ab} of a wire from a to b is the reciprocal of the resistance R_{ab}

$$G_{ab} = R_{ab}^{-1} = \frac{I_{ab}}{V_{ab}} \tag{3.48}$$

The unit of conductance is Ω^{-1} or S, siemens.

If we cut or remove the wire, we get a zero value for the conductance, $G = 0$. If we replace the wire by a superconducting wire,[12] we get an infinitely high conductance $G \to \infty$. The conductance of a "normal" wire (not a superconducting wire) is proportional to the cross-sectional area A and inversely proportional to the length L of the wire. This may be expressed in the following equation

$$G = R^{-1} = \rho^{-1}\,\frac{A}{L} = \sigma\,\frac{A}{L} \tag{3.49}$$

The proportionality constant σ (greek letter "sigma") is the *conductivity*. The conductivity is the reciprocal of the resistivity

$$\sigma = \rho^{-1} \tag{3.50}$$

If we substitute Eq. 3.50 into Eq. 3.47 we find the following equation.

$$R = \frac{L}{\sigma A} \tag{3.51}$$

The conductivity is, like the resistivity, independent of the length and independent of the cross-sectional area of the wire. The conductivity depends only on the material and the temperature.

[12] We assume a constant DC current which is less than the critical current.

3.5 Passive and Active Configuration of a Circuit Element (Method 1)

Before starting to calculate the voltages and currents in a circuit, we should define a positive voltage and a positive current direction for each circuit element, see Fig. 3.10. Configuration of a circuit element means that we assign the circuit element a positive orientation for voltage and current. We do this as follows:

1. We draw an arrow to define a positive current direction for the circuit element. If a positive charge flows in this direction (or a negative charge flows in the opposite direction), the current I is positive.
2. We mark the two terminals of the circuit element with a plus and a minus sign. The voltage V is the difference in potential between the terminal with a plus sign and the terminal with a minus sign. Thus, voltage V is the potential drop from the plus-marked terminal to the minus-marked terminal

$$V \equiv V_+ - V_- \tag{3.52}$$

If the potential is higher at the plus marked terminal than at the minus marked terminal, the voltage V is positive.

3.5.1 The Passive Configuration

Figure 3.10 shows the passive configuration. It is typically used for a passive circuit element (such as resistance) where we have a potential drop in the current direction.

Theorem 9 ***The passive configuration*** *is defined by the positive current direction pointing toward the terminal marked with a plus sign. This configuration is used mainly for passive circuit elements, e.g. resistors.*

If both V and I are positive, we see that the positive charge dq enters the circuit element at the terminal marked with a plus sign with the highest potential. Now, dq flows through the circuit element from plus to minus in the positive current direction. When dq comes out of the resistive circuit element at the terminal marked minus, it has a lower potential and lower energy than before at the terminal marked plus. There has been a potential drop, which means that dq has lost energy. If the circuit element is a resistor, then the energy loss is converted to heat in the resistor. If P is positive in a passive configuration, there is a flow of power into the circuit element. The power loss is

$$P \equiv V \cdot I \tag{3.53}$$

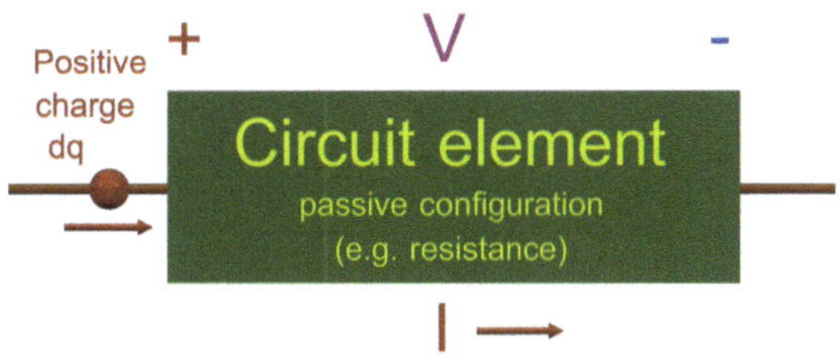

Fig. 3.10 The passive configuration of a circuit element. N.E.Dam ©

3.5.2 The Active Configuration

Figure 3.11 shows the active configuration. It is typically used for an active circuit element (like a battery), where we have an increase in potential in the current direction.

We draw an arrow to indicate the current direction. If a positive charge flows in this direction (or a negative charge flows in the opposite direction), the current I is positive.

We assign a plus and a minus sign for the voltage. If the potential is higher at the plus-marked terminal than at the minus marked terminal, the voltage V is positive.

Theorem 10 ***The active configuration*** *is defined by the positive current direction pointing away from the terminal marked with a plus sign. This configuration is mainly used for active circuit elements, e.g. batteries.*

If both V and I are positive, we see that the positive charge dq enters the circuit element at the terminal marked minus with the lowest potential. Now, dq flows through the circuit element from minus to plus in the positive current direction. When dq leaves the circuit element at the plus-marked terminal, it has a higher potential than before at the minus-marked terminal and therefore it has a higher energy. In order to indicate that the circuit element develops energy and increases the energy of dq , the formula for power is written in the following way, where the negative sign for power means that the circuit element supplies energy to the circuit connected to the circuit element.

$$P \equiv -V \cdot I \tag{3.54}$$

3.6 General Configuration of a Circuit Element (Method 2)

In Method 1, we operate with two different types of configurations and two different expressions for the power. In Method 2, we operate with only one type of configuration and only one expression for the power. We would prefer to use Method 2 in this textbook.

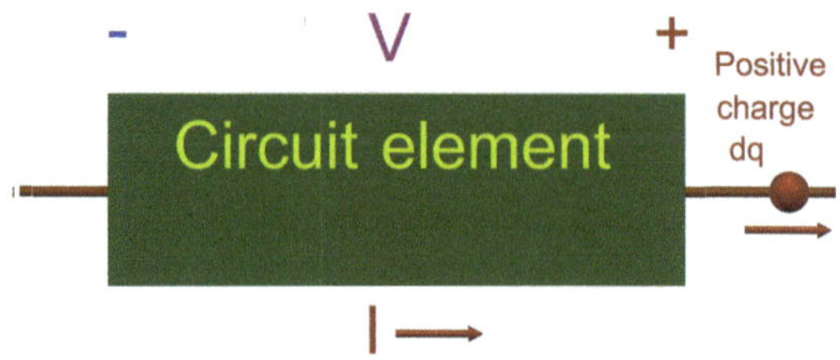

Fig. 3.11 The active configuration of a circuit element. N.E.Dam ©

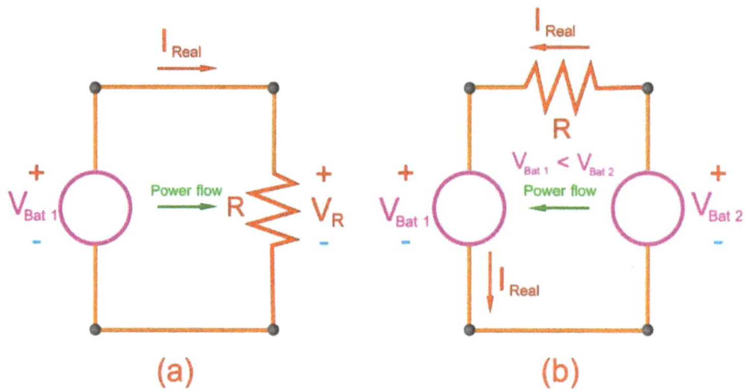

Fig. 3.12 The Method 2 configuration of circuit elements. N.E.Dam ©

Theorem 11 ***The Method 2 configuration*** *uses the same configuration and the same expression for the power for both passive and active circuit elements. It is defined this way:*

- *The reference direction for the positive current is away from the plus terminal. This current is denoted by* I_+.
- *The current* I_+ *is considered positive if the real current flow is away from the plus terminal.*
- *The current* I_+ *is considered negative if the real current flow is into the plus terminal.*
- *The power associated with the circuit element is always* $\mathrm{P} = \mathrm{V} \cdot \mathrm{I}_+$.

In Fig. 3.12a we have a battery with voltage V_{Bat1}. The plus and minus poles of the battery are marked with the red plus sign and the blue minus sign.
The real physical current flows in the direction shown with the arrow below I_{Real}.
In addition, we have a resistor with voltage V_R (=V_{Bat1}). When current flows through the resistor, a drop in potential occurs in the current direction, i.e. we have the highest potential at the top of the resistor (marked with a red plus sign).

The power associated with the battery

As the current flows out of the plus pole of the battery, the current I_+ is positive. At the same time, the voltage is positive. The power therefore becomes

$$P = V \cdot I_+ = V_{Bat1} \cdot I_{Real} > 0 \tag{3.55}$$

A positive power means that the circuit element is delivering power, i.e. there is a power flow out of the battery to the right.[13]

The power associated with the resistor

The current I_{Real} flows into the resistor at the resistor's plus terminal, where the potential is highest. The current I_+ away from the resistor's plus terminal therefore becomes

$$I_+ = -I_{Real} \tag{3.56}$$

Since the voltage is positive, we find the power associated with the resistor

$$P = V \cdot I_+ = V_R \cdot (-I_{Real}) < 0 \tag{3.57}$$

The fact that there is a negative power associated with the resistor means that the resistor absorbs power. There is therefore a power flow into the resistor (towards the right in the figure).[14]

In Fig. 3.12b we have a battery with voltage V_{Bat1} and another battery with voltage V_{Bat2}. Since V_{Bat2} is larger than V_{Bat1}, there is a higher potential at the positive pole of Battery 2 than at the positive pole of Battery 1. Therefore, the actual direction of the current is as shown by the arrow next to I_{Real}. The current flows in the opposite direction than it did in figure a and flows in the opposite direction through Battery 1, which means that Battery 1 is charged.[15]

The power associated with Battery 1

The current I_{Real} flows into the plus terminal of Battery 1. The current I_+ away from the plus terminal of Battery 1 therefore becomes

$$I_+ = -I_{Real} \tag{3.58}$$

---

[13] NB! With Method 1 we have (per definition) a negative power for a circuit element that emits power to the surroundings. With Method 2 we have a positive power for a circuit element that emits power to the surroundings.

[14] NB! With Method 1 we have (per definition) a positive power for a circuit element that absorbs power from the surroundings. With Method 2 we have a negative power for a circuit element that absorbs power from the surroundings.

[15] We assume that Battery 1 is a rechargeable battery.

Since the voltage is positive, we find the power associated with Battery 1

$$P = V \cdot I_+ = V_{Bat1} \cdot (-I_{Real}) < 0 \tag{3.59}$$

The fact that there is a negative power associated with Battery 1 means that Battery 1 absorbs power. There is therefore a power flow into Battery 1 (towards the left in the figure).

The power associated with Battery 2
As the current flows out of the plus pole of Battery 2, the current is positive. At the same time, the voltage is positive. The power therefore becomes

$$P = V \cdot I_+ = V_{Bat2} \cdot I_{Real} > 0 \tag{3.60}$$

A positive power means that Battery 2 is delivering power, i.e. there is a power flow out of the battery to the left.

The charging current
The potential drop in the direction of current, i.e. the voltage, across the resistor R is given by

$$V_R = V_{Bat2} - V_{Bat1} \tag{3.61}$$

Thus, according to Ohm's law, the current becomes

$$I_{Real} = \frac{V_R}{R} = \frac{V_{Bat2} - V_{Bat1}}{R} \tag{3.62}$$

The charging current is therefore limited by the resistance R. If $R = 0$ the current becomes, in principle, infinitely large, corresponding to the two batteries being short-circuited. In that case, the current is limited by only any internal resistance in the two batteries, but even this current can be destructive to the two batteries. The internal resistance of a battery is described in more detail in the Sect. 3.9 on page 80.

3.7 Kirchhoff's Current Law

We repeat the previously derived Kirchhoff's Current Law

Theorem 12 ***Kirchhoff's Current Law (KCL).*** *The sum of currents in a node is equal to the sum of currents away from the same node.*

$$\Sigma_i I_i \text{ in a node} = \Sigma_j I_j \text{ away from the same node}$$

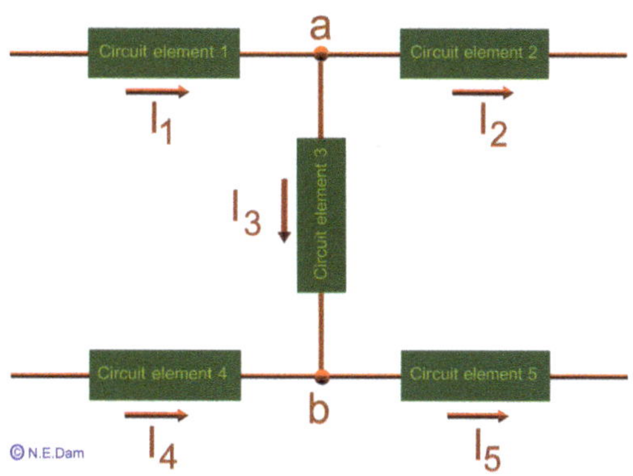

Fig. 3.13 A part of a circuit with two nodes, a and b, used to illustrate KCL

Theorem 12 is one of the most fundamental theorems in electrical circuit theory.

Figure 3.13 shows a part of a circuit with two nodes, a and b. If we consider Kirchhoffs current law (KCL) for node a, we find the following.

$$I_1 = I_2 + I_3 \tag{3.63}$$

This expresses that charge cannot accumulate at node a. If node a is a capacitor plate, it is possible that charge can accumulate at node a. In this case we could have a time interval, where more positive charge flows into node a than away from node a:

$$I_1 > I_2 + I_3 \tag{3.64}$$

In this case KCL is not valid!

3.8 Kirchhoff's Voltage Law

We repeat the previously derived Kirchhoff's voltage Law

Theorem 13 ***Kirchhoff's voltage Law (KVL)*** *states that if we move around a closed electrical loop, the sum of potential increases will equal the sum of potential decreases.*

$$\Sigma_i V_i \text{ (potential increases in a loop)} = \Sigma_j V_j \text{ (potential decreases in a loop)}$$

Figure 3.14 shows a part of a circuit with two loops, 1 and 2. Consider loop 1 as an example to illustrate the Kirchhoff voltage law (KVL), and let us follow a positive charge moving in the loop direction. If a charge moves around in the shown direction in loop 1 we find that

$$V_1 = V_2 + V_3 \tag{3.65}$$

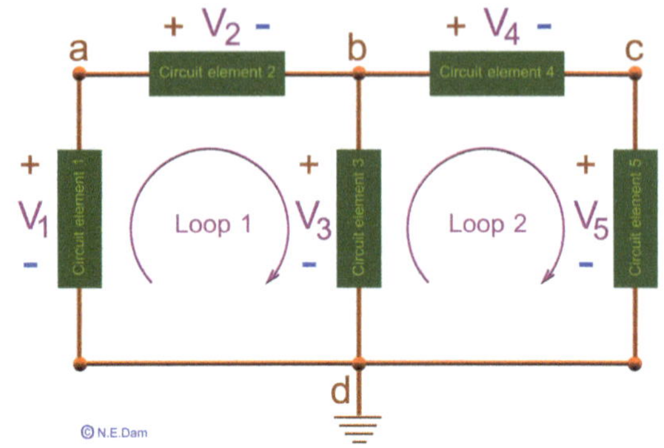

Fig. 3.14 A part of a circuit with two loops, 1 and 2, used to illustrate KVL

When we "move" in loop 1, we have the following: By passing through element 1 of the circuit from − to + in the loop direction, we have an increase in potential, which is V_1. Passing through circuit element 2 from + to − in the loop direction, we have a decrease in potential, which is V_2. Passing through circuit element 3 from + to − in the loop direction, we have a decrease in potential, which is V_3. Thus, we find Eq. 3.65.

The following theorems are the three most fundamental laws in electrical circuit theory

- Ohm's law Theorem 8 on page 71
- Kirchhoff's Current Law Theorem 12 on page 78
- Kirchhoff's voltage Law Theorem 13 on page 79

3.9 Voltage Sources

A source—or generator—is a circuit element that produces energy to the circuit and creates currents in the circuit. A well-known example of a DC voltage source is a battery. A battery can—to some extent—be approximated by an equivalent electric circuit as shown in Fig. 3.15. When a positive charge flows through a source from the minus pole to the plus pole, the potential is increased (from a to b). When the charge then flows through passive circuit elements such as resistances, the potential drops in steps (from b to c through the internal resistance and from c to d through the external resistance), eventually the potential becomes zero (at the point d) before it circulates back to the source at the point a.

Figure 3.16 shows a general voltage source with voltage V_s. The *internal resistance* of the source, R_i causes a decrease in the output voltage if the output current $I > 0$. Thereby, the output voltage becomes:

$$V = V_s - R_i I \tag{3.66}$$

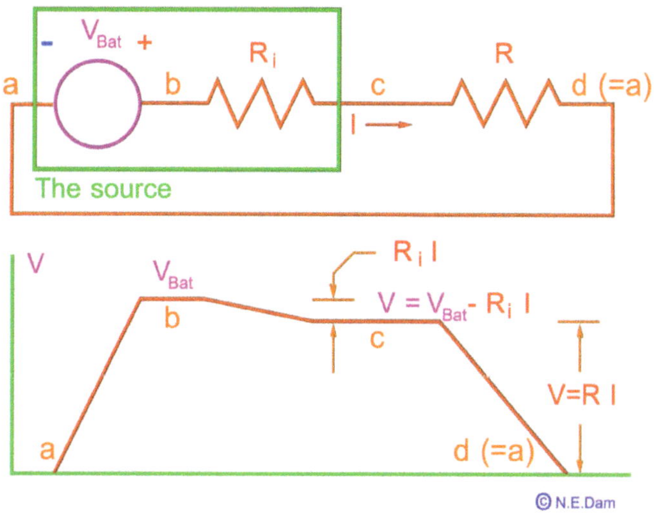

Fig. 3.15 A battery in a circuit with an internal and external resistance, identical with Fig. 3.8.

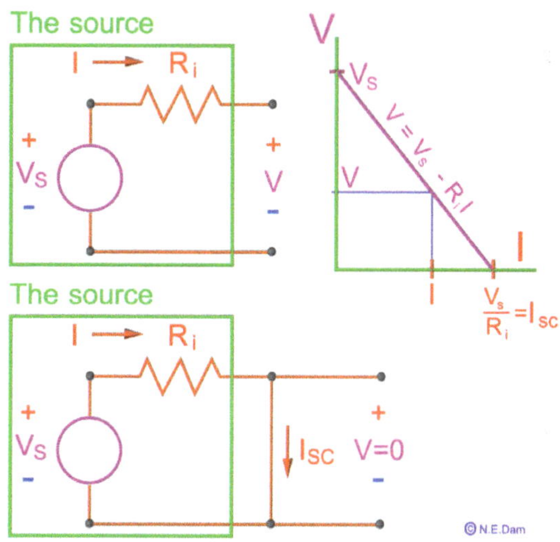

Fig. 3.16 A general voltage source and the output voltage V as a function of the current I.

To obtain a variable current I in the circuit, it is necessary to place a variable resistor between the two output terminals (not shown in Fig. 3.16, but shown as the resistor R in Fig. 3.15). Figure 3.16 shows a representation of Eq. 3.66. It shows that the output voltage V is a linear function of the current I through the source. This depiction of the output voltage V as a function of the output current I is denoted as the *output characteristic*.

Open circuit

An *open circuit* means that we do not have a load on the output terminals. If $I = 0$ there is no potential drop across the internal resistance R_i. According to Eq. 3.66, the output voltage becomes $V = V_s$, which is the battery's output voltage without load, the open circuit voltage. This is the point $(I, V) = (0, V_s)$ where the straight line intersects the vertical axis.

Short circuit

Now, we make a *short circuit* through the output terminals, as illustrated in the lower part of Fig. 3.16. The current is now the short circuit current I_{sc} , which is only limited by the internal resistance R_i. If we assume that there is no potential drop in the short circuit, then $V = 0$ can be substituted into Eq. 3.66 and we find $I_{sc} = V_s/R_i$. Then we have the point $(I, V) = (I_{sc}, 0)$ where the straight line intersects the horizontal axis. We could use these two points to find the straight line with slope $-R_i$ as illustrated in Fig. 3.16. In practice, it is not recommended to short circuit the source! It is better to measure the output voltage V for another—not harmful—current value I and then use this point (I, V) as the second point needed to estimate the straight line.

An ideal voltage source

An *ideal voltage source* has zero internal resistance, $R_i = 0$. This means that the output voltage V versus I in Fig. 3.16 becomes a straight horizontal line with slope $-R_i = 0$. *The output voltage V is then constant and independent of the current I that flows out of the output terminals*. The diagram symbol can be a circle with the voltage V_s written next to it, as shown in Fig. 3.17a. If the internal resistance $R_i = 0$, the entire green box is an ideal voltage source.

A real voltage source

A *real voltage source* has an internal resistance $R_i > 0$ as the green box shown in Fig. 3.17a. This means that the output voltage V versus I in Fig. 3.16 becomes a straight line with a negative slope, i.e., $-R_i < 0$. *The output voltage V is a decreasing function of the current I that flows from the output terminals*; see Fig. 3.16.

An independent voltage source

Figure 3.17a shows an *independent voltage source*. The circle in the green box is the ideal voltage source with the voltage V_s. The term "ideal" means, as mentioned, that V_s is independent of how much current flows through it. The term "independent" means that V_s does not depend on other voltages or currents.

A dependent voltage source

Figure 3.17b shows a voltage *dependent voltage source*, where V_s depends on another voltage V_3. The constant α is dimensionless. The voltage V_3 can be another voltage in the same electrical circuit as V_s.

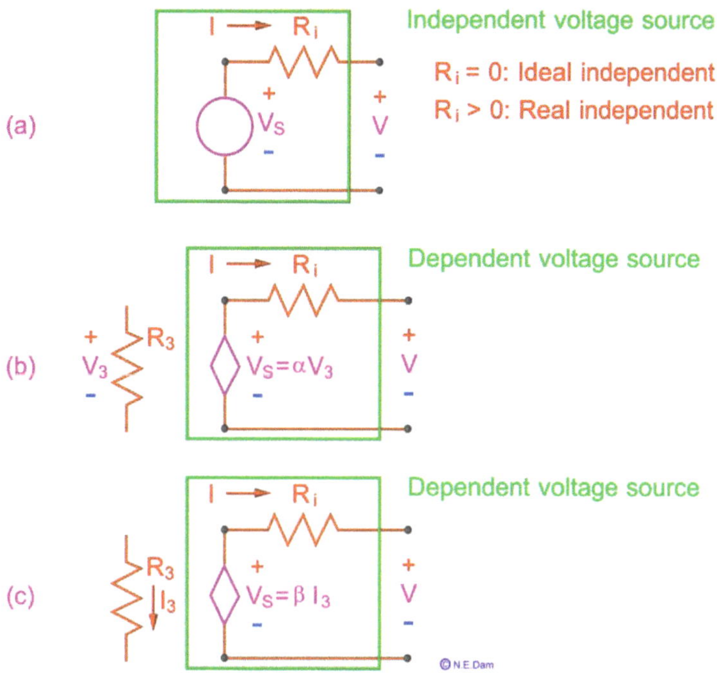

Fig. 3.17 One independent and two dependent voltage sources. R_3 in figures (b) and (c) is drawn outside the green box, as it may belong to a different electrical circuit than the circuit in the green box. In that case, the information about V_3 or I_3 may be transferred to V_s via a magnetic or optical circuit

V_3 can also be a voltage in another electrical circuit that is "*galvanically separated*" from V_s, i.e., there is no electrical connection between the two circuits. In that case, there can be a transformer, i.e. a connection via a magnetic circuit between the two electrical circuits. There can also be an *optocoupler*,[16] i.e. an optical connection between the two circuits. In short, the information about V_3 is transferred somehow to V_s without a direct electrical connection. If V_s depends on a current I_3, as in Fig. 3.17c, the constant β has the dimension volt/ampere.

[16] An optocoupler is a component that contains a light-emitting diode (LED) and a photosensitive diode that are not electrically connected; thus, it is possible to separate two electrical circuits but still transmit information via light pulses.

3.10 Current Sources

Figure 3.18a shows an independent current source. Usually, the current source is drawn with the internal conductance in parallel with the source. The internal conductance is the reciprocal of the internal resistance[17]

$$G_i = \frac{1}{R_i} \tag{3.67}$$

Using KCL at the upper node, we find

$$I_s = I_i + I \Rightarrow I_s R_i = I_i R_i + I R_i = V + I R_i \Rightarrow V = I_s R_i - I R_i \tag{3.68}$$

If we divide this equation by R_i then

$$V = I_s R_i - I R_i \Rightarrow \frac{V}{R_i} = I_s - I \Rightarrow I = I_s - G_i V \tag{3.69}$$

This is the equation for IV characteristic in Fig. 3.18. Please note the analogy to the VI characteristic in Fig. 3.16, if V is swapped with I and R_i with G_i.

If we short circuit the output of the current generator in Fig. 3.18a, then we can ignore the current I_i through G_i, since G_i is in a parallel connection with a short circuit with conductance $G_{sc} \to \infty$. The current in the short circuit therefore becomes $I_{sc} = I_s$.

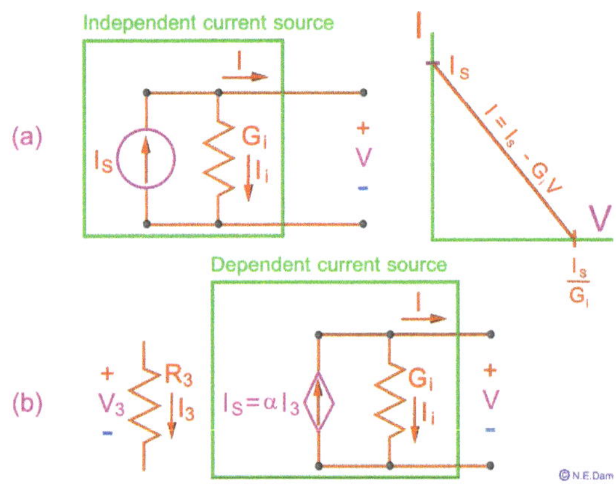

Fig. 3.18 An independent and a dependent current source

[17] It does not matter whether we use the internal resistance R_i or the internal conductance G_i. They have the same circuit element symbol (a resistance), which - in both cases - should be placed in parallel with the source I_s. E.g., if the internal resistance $R_i = 2\ \Omega$, we write $R_i = 2\ \Omega$ or just R_i beside the circuit element symbol (the resistance). We could also write $G_i = 2^{-1}\Omega^{-1} = 0,5\ \Omega^{-1}$ or just G_i beside the same circuit element symbol (the resistance).

Comparison between the current generator and the voltage generator
If we assume that the current generator has the same short circuit current as the voltage generator in Fig. 3.16, we find that

$$I_s = I_{sc} = \frac{V_s}{R_i} \tag{3.70}$$

If we substitute this into Eq. 3.68 we find the following.

$$I_s = I_{sc} = \frac{V_s}{R_i} \Rightarrow V = \frac{V_s}{R_i} R_i - I R_i \Rightarrow V = V_s - R_i I \tag{3.71}$$

This result is identical with the result for the voltage output from the voltage source in Fig. 3.16 and Eq. 3.66. This means that we can use the circuit in Fig. 3.18a as an approximate model of a source or a battery, if the current source has the following current and the following internal resistance:

$$I_s = I_{sc} = \frac{V_s}{R_i} \Rightarrow R_i = \frac{V_s}{I_{sc}} \Rightarrow G_i = \frac{I_{sc}}{V_s} \tag{3.72}$$

Please note that the internal resistance R_i (or the internal conductance G_i) in Fig. 3.18a is in a parallel connection with the source I_s and also in a parallel connection with the output voltage V given by Eq. 3.70, which is identical to Eq. 3.66.

An ideal current source
Figure 3.18a shows an *ideal current source* if $R_i = \infty \Leftrightarrow G_i = 0$. The IV characteristic in Fig. 3.18a therefore becomes a horizontal line $I = I_s$.

A real current source
Figure 3.18a shows a *real current source* if $G_i > 0$. The IV characteristic in Fig. 3.18a therefore becomes a line with a negative slope $-G_i < 0$.

An independent current source
Figure 3.18a shows an independent current source. The current I_s is independent of other currents and other voltages. The concept of "independent" is explained in the previous paragraph, "An independent voltage source".

A dependent current source
Figure 3.18b shows a dependent current source. The current I_s depends on a voltage or on another current. If I_s depends on another current, the constant α is dimensionless. If I_s depends on a voltage, the constant α has the dimension ampere/volt. The concept of "dependent" is explained in the previous paragraph "A dependent voltage source".

3.11 Tests

General guidelines for answering the test
The test is without any aids; only pencils, rulers, etc. are permitted. The evaluation of the test, which is handed in, reflects not only the correctness of the answer but also the care and dedication that has been put into the answer.

A. Write the answer or formula correctly and show all the steps in the calculations.
B. The answer must be well structured, orderly, and easy to read.
C. Briefly explain the symbols used the first time they appear in the test.
D. If suggested, make a small sketch or diagram in colour to add clarity.

3.11.1 Test 3.1 Basic Definitions with Circuit Elements

Problem 1 (7%) Electric current

A. Explain the definition of electric current, possibly using a small sketch.
B. Let $q(t) = 0,4 + 0,8\,t$. What is $I(t)$?
C. Let $q(t) = 0,8\,cos(\omega t + \phi)$, where ω and ϕ are constants. What is $I(t)$?

Problem 2 (20%) voltage

A. Write down the electric potential energy, $U_{epot,a}$, of a charge q_o relative to the three charges q_1, q_2, q_3, when the charge q_o is located at the point a, see Fig. 3.19.
B. Write down the electric potential energy, $U_{epot,b}$, of a charge q_o relative to the three charges q_4, q_5, q_6, when the charge q_o is located at the point b, see Fig. 3.19.

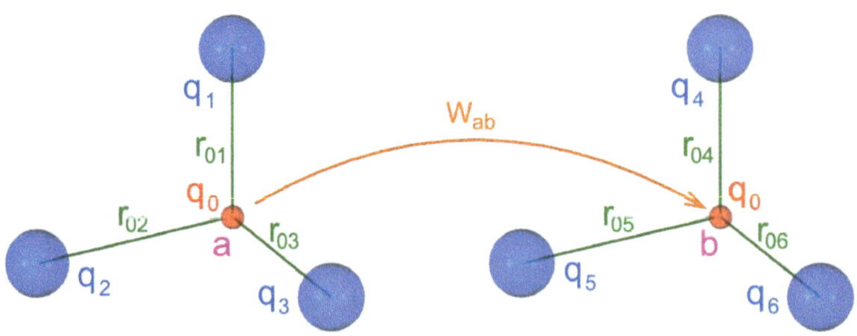

Fig. 3.19 The figure is used when calculating the electric potential energy of a charge q_o relative to the three charges q_1, q_2, q_3, when the charge q_o is located at the point a. It is assumed that q_o is so far away from the charges q_4, q_5, q_6 that we may neglect the electric potential energy that q_o obtain relative to q_4, q_5, q_6. N.E.Dam ©

C. How much work W_{ab} is required to move q_o from point a to point b[18]?
D. What is the drop in potential V_{ab} when q_o moves from point a to point b[19]?
E. When we replace q_o with an infinitesimal charge dq, the work W_{ab} becomes dW_{ab}. What is the definition of an electrostatic voltage?

Problem 3 (7%) Power

A. Write down and explain the definition of electric power P.
B. Multiply the definition of electric power by $\frac{dq}{dq}$ and identify the electric current and the electrostatic voltage.
C. Write down the electric power expressed by the electric current and the electrostatic voltage.

Problem 4 (14%) Resistance

A. Write down the current-voltage relationship for a resistor and state the condition for the relationship to be Ohm's law.
B. Sketch $V(I)$ for an Ohmic material and a non-Ohmic material.
C. Can the current-voltage relationship be used if the resistance depends on the current?
D. If resistance depends on current, define the static and dynamic resistance at a point (I_1, V_1) on the curve $V(I)$.
E. Write down the expression for resistance on a wire with cross-sectional area A and length L and explain the proportionality factor.
F. What is conductance and what is conductivity? Write down the expression for the resistance of a wire, where we use conductivity instead of resistivity.

Problem 5 (7%) Passive configuration of a circuit element, Method 1

A. Make a sketch showing the passive configuration. Explain when voltage and current are positive.
B. Write down the formula for calculating the power P. What does it mean physically if P is positive or negative?

Problem 6 (7%) Active configuration of a circuit element, Method 1

A. Make a sketch showing the active configuration. Explain when voltage and current are positive.
B. Write down the formula for calculating the power P. What does it mean physically if P is positive or negative?

---

[18] Hint: Eq. 3.13.
[19] Hint: Eq. 3.17.

Problem 7 (7%) General configuration of a circuit element (Method 2)

A. Make a sketch showing the general configuration. Explain the condition for which the current I_+ is positive.
B. Write down the formula for calculating the power P using V and I_+. What does it mean physically if P is positive or negative?

Problem 8 (14%) Kirchhoff's two laws

A. Make a sketch with a simple example and formulate Kirchhoff's Current Law in words.
B. Make a sketch with a simple example and formulate Kirchhoff's Voltage Law in words.

Problem 9 (10%) Voltage source

A. Make a sketch of a voltage source V_s with internal resistance R_i.
B. Describe how to find the voltage V in the output and draw the output characteristic of $V(I)$.
C. What is an ideal voltage source?
D. What is an independent voltage source?

Problem 10 (7%) Current source

A. Make a sketch of a current source I_s in parallel with $G_i = 1/R_i$.
B. Describe how to find the current I in the output and draw the output characteristic of $I(V)$.

Resistive Circuits

4

4.1 Introduction

In this chapter, we describe circuits with resistance, DC voltage generators, and DC current generators. We deal with series and parallel connexions, voltage and current dividers, Thévenin and Norton equivalents. Finally, we use the Node voltage method and the Mesh current method to solve more complicated circuits. The methods described in this chapter are, in fact, valid for resistive circuits but can also be used for circuits with capacitors and inductances.

4.2 Series Connection of Resistances

A series of resistances is shown in the upper part of Fig. 4.1. All resistances have a passive configuration with a positive current direction in the plus sign for the voltage. For each of the circuit elements the Ohm's law may be used in this way

$$V_1 = R_1 I, \; V_2 = R_2 I, \; V_3 = R_3\, I \tag{4.1}$$

Now, we use KVL around in the upper circuit in the current direction. Going through the voltage generator from − to + we have an increase in potential, which is V. This equals the total drop in potential from a to d in the three resistances

$$V = V_1 + V_2 + V_3 = (R_1 + R_2 + R_3)\, I = R_{eq}\, I \tag{4.2}$$

Hence, the series connection acts like an equivalent resistance with the value

$$\boldsymbol{R_{eq} = R_1 + R_2 + R_3} \tag{4.3}$$

N. Dam, *An Introduction to Electrical Circuit Theory*, Synthesis Lectures on Engineering, Science, and Technology, https://doi.org/10.1007/978-3-032-34346-8_4

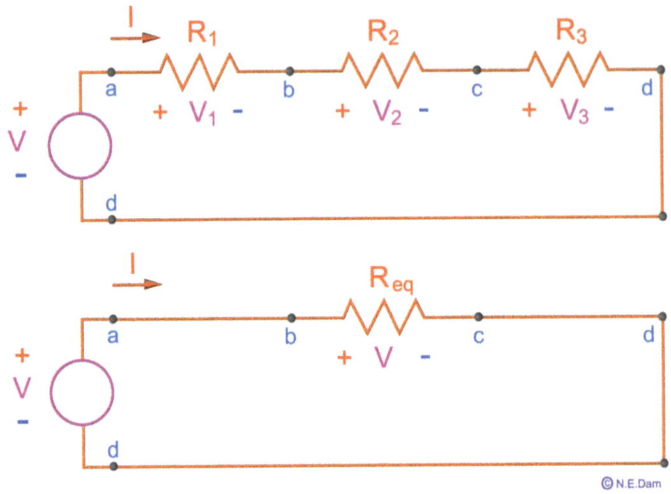

Fig. 4.1 Series connection of resistances

Equation 4.3 can be expanded to more than three terms.

4.3 Parallel Connection of Resistances

A parallel connection is shown in Fig. 4.2. We have the same potential drop, V , in all three resistances and we have a passive configuration with a positive current direction in the plus sign for the voltage for each of the three resistances.

Fig. 4.2 Parallel connection of resistances

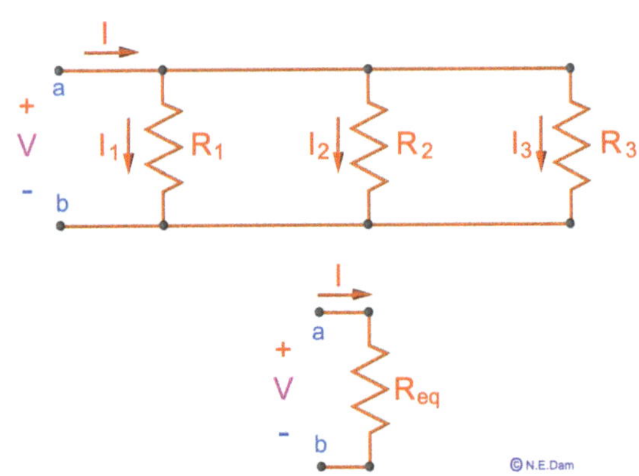

For each of the circuit elements the Ohms law may be used in the following way:

$$I_1 = \frac{V}{R_1},\ I_2 = \frac{V}{R_2},\ I_3 = \frac{V}{R_3} \tag{4.4}$$

The upper nodes are common for the three resistances and may be considered as a single node. Using KCL for this node, we get the following.

$$I = I_1 + I_2 + I_3 = \left(\frac{1}{R_1} + \frac{1}{R_2} + \frac{1}{R_3}\right) V = \frac{1}{R_{eq}} V \tag{4.5}$$

Hence, parallel connection acts as an equivalent resistance with a value R_{eq} that may be determined by the equation

$$\boldsymbol{\frac{1}{R_{eq}} = \frac{1}{R_1} + \frac{1}{R_2} + \frac{1}{R_3}} \tag{4.6}$$

If we use conductances instead of resistances, we end up with

$$\boldsymbol{G_{eq} = G_1 + G_2 + G_3} \tag{4.7}$$

It is obvious that Eqs. 4.6 and 4.7 can be expanded to n terms

$$\boldsymbol{\frac{1}{R_{eq}} = \frac{1}{R_1} + \frac{1}{R_2} + \cdots + \frac{1}{R_n}} \tag{4.8}$$

If we use conductances instead of resistances, we end up with

$$\boldsymbol{G_{eq} = G_1 + G_2 + \cdots + G_n} \tag{4.9}$$

In practice, if we have more than two resistances in a parallel connection, it may be easier to change the resistances to conductances first and then add up all the conductances to get a total conductance. The total conductance may eventually be converted to a resistance if needed.

If we have a parallel connection with only two resistances, we find

$$\frac{1}{R_{eq}} = \frac{1}{R_1} + \frac{1}{R_2} = \frac{R_2}{R_1 R_2} + \frac{R_1}{R_1 R_2} = \frac{R_1 + R_2}{R_1 R_2} \tag{4.10}$$

Hence, for a parallel connection of two resistances, we find

$$\boldsymbol{R_{eq} = \frac{R_1 R_2}{R_1 + R_2}} \tag{4.11}$$

If $R_1 = R_2$ for two resistances in a parallel connection, the equivalent resistance simply becomes

$$R_{eq} = \frac{R_1 R_1}{2R_1} = \frac{R_1}{2} \tag{4.12}$$

4.4 Voltage Divider

A voltage divider is shown in Fig. 4.3. All resistances have a passive configuration with the positive current direction in the plus sign for the voltage. We use KVL around the circuit in the current direction. Going through the voltage generator from - to + we have an increase in potential, which is V_S. This equals the total drop in potential from a to d in the three resistances.

$$V_S = V_1 + V_2 + V_3 = (R_1 + R_2 + R_3)\, I \tag{4.13}$$

Hence, the current has the value

$$I = \frac{V_S}{R_1 + R_2 + R_3}$$

The drop in potential across R_1 is

$$V_1 = R_1\, I = V_S\, \frac{R_1}{R_1 + R_2 + R_3} \tag{4.14}$$

or

$$\frac{\boldsymbol{V_1}}{\boldsymbol{V_S}} = \frac{\boldsymbol{V_1}}{\boldsymbol{V_1 + V_2 + V_3}} = \frac{\boldsymbol{R_1}}{\boldsymbol{R_1 + R_2 + R_3}} \tag{4.15}$$

This result may be expressed as follows:

V_1 relative to the total potential drop across the resistances, $\mathrm{V_S = V_1 + V_2 + V_3}$ *is the same ratio as* $\mathrm{R_1}$ *relative to the total resistance* $\mathrm{R_1 + R_2 + R_3}$.

A similar result may be derived for the second and third resistances

$$\frac{V_2}{V_S} = \frac{V_2}{V_1 + V_2 + V_3} = \frac{R_2}{R_1 + R_2 + R_3} \tag{4.16}$$

and,

$$\frac{V_3}{V_S} = \frac{V_3}{V_1 + V_2 + V_3} = \frac{R_3}{R_1 + R_2 + R_3} \tag{4.17}$$

The results may be expanded to have more resistances than the three shown here.

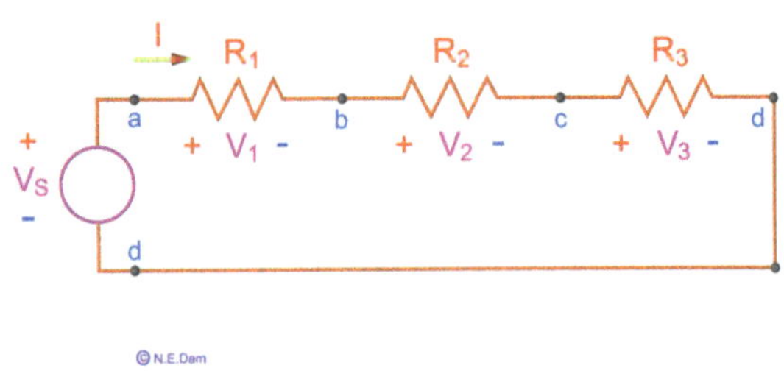

Fig. 4.3 Voltage divider

In the case where there are only two resistances, Eq. 4.15 becomes

$$\frac{V_1}{V_S} = \frac{V_1}{V_1 + V_2} = \frac{R_1}{R_1 + R_2} \tag{4.18}$$

or

$$\boldsymbol{V_1 = V_S \frac{R_1}{R_1 + R_2}} \tag{4.19}$$

4.5 Current Divider

A current divider is shown in Fig. 4.4. All conductances have a passive configuration, with the positive current direction in the plus sign for the voltage. The parallel connection of the conductances has the following total equivalent conductance

$$G_{eq} = G_1 + G_2 + G_3 \tag{4.20}$$

The potential drop across this parallel connection of conductances is

$$V = R_{eq}\, I_S = \frac{I_S}{G_{eq}}$$

Now, the current through R_1 is

$$I_1 = \frac{V}{R_1} = V\, G_1 = \frac{I_S}{G_{eq}}\, G_1 = I_S \frac{G_1}{G_{eq}} = I_S \frac{G_1}{G_1 + G_2 + G_3}$$

The current divider formula becomes

$$\boldsymbol{\frac{I_1}{I_S} = \frac{I_1}{I_1 + I_2 + I_3} = \frac{G_1}{G_1 + G_2 + G_3}} \tag{4.21}$$

This result may be expressed as follows:
I_1 *relative to the total current to the conductances,* $I_S = I_1 + I_2 + I_3$ *is the same ratio as* G_1 *relative to the total conductance* $G_1 + G_2 + G_3$.

If we compare with the formula for the voltage divider, Eq. 4.15:

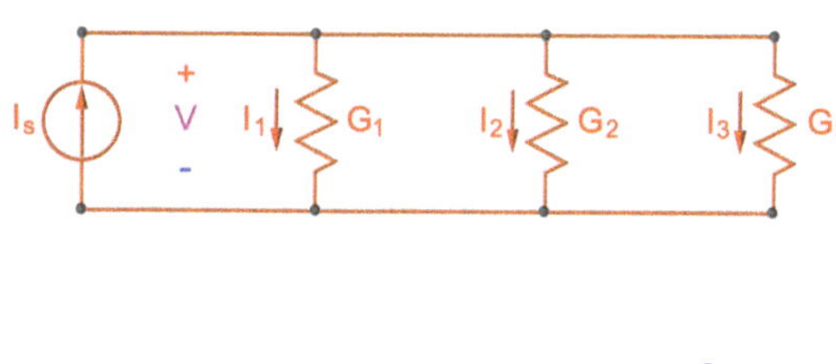

Fig. 4.4 Current divider

$$\frac{V_1}{V_S} = \frac{V_1}{V_1 + V_2 + V_3} = \frac{R_1}{R_1 + R_2 + R_3}$$

we see an analogy between the two formulas. If we remember the formula for the voltage divider, we can transform the voltage divider formula into a current divider formula as follows:

- The series connection of the resistances becomes a parallel connection of the conductances.
- R_1, R_2, R_3 in the voltage divider formula are replaced by G_1, G_2, G_3
- V_S, V_1, V_2, V_3 in the voltage divider formula are replaced by I_S, I_1, I_2, I_3

Similar formulas may be derived for the second and third conductances.

$$\frac{I_2}{I_S} = \frac{I_2}{I_1 + I_2 + I_3} = \frac{G_2}{G_1 + G_2 + G_3} \tag{4.22}$$

and

$$\frac{I_3}{I_S} = \frac{I_3}{I_1 + I_2 + I_3} = \frac{G_3}{G_1 + G_2 + G_3} \tag{4.23}$$

The results may be expanded to more conductances than the three shown here. In the case where there are only two conductances, Eq. 4.21 becomes

$$\frac{I_1}{I_S} = \frac{I_1}{I_1 + I_2} = \frac{G_1}{G_1 + G_2} \tag{4.24}$$

or

$$\boldsymbol{I_1 = I_S \frac{G_1}{G_1 + G_2}} \tag{4.25}$$

4.6 Thévenin Equivalent Circuit

The green box in Fig. 4.5 is considered a two-terminal circuit with resistances together with voltage and/or current generators. If there are dependent generators in the circuit, they should depend on voltages and currents inside the circuit, i.e. inside the green box.

The two terminal circuit in the green box may be replaced by a simple equivalent circuit, the Thévenin equivalent circuit. If we look at the green box from outside, the equivalent circuit has the same properties, i.e., it has the same voltage V across the output terminals a and b, and the current flowing through the terminals will be the same.

Practical use of the Thévenin equivalent circuit

When we replace the green box on the left side of Fig. 4.5 with its Thévenin equivalent circuit on the right side of the figure, it becomes much easier to calculate currents and voltages in an external circuit that is connected. In general, we are not interested in detailed information

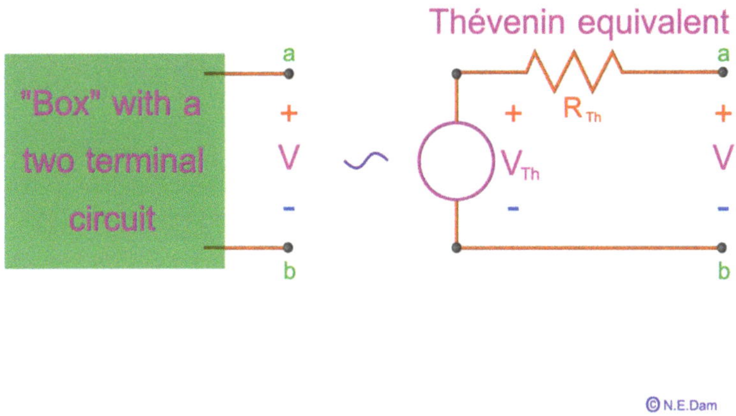

Fig. 4.5 The Thévenin equivalent circuit

about the currents and voltages inside the green box. The output terminals (a and b) of the green box could, for example, be an electrical outlet in an apartment. The circuit in the green box could then be all the wires, transformers, etc. between the electrical outlet in the apartment and the power plant. If we connect an apparatus to the electrical outlet and want to calculate the current in the wire to the apparatus, then it would be very complicated if we had to calculate all the currents in all the wires on the way to the power plant. In this case, we are not interested in what happens in the green box, which represents the circuits from the electrical outlet to the power plant. We are only interested in the output terminal voltage and current, and these quantities are easier to calculate when the green box is replaced by its equivalent Thévenin circuit on the right side of the Fig. 4.5. If we did not replace the green box, we would need detailed information about the circuit inside the green box, and it could be very complicated to calculate the output terminal voltage and the output current if we should take into account the detailed circuit in the green box.

The Thévenin equivalent circuit was discovered by Hermann von Helmholtz in 1853 and again independently in 1883 by the French telegraph engineer Léon Charles Thévenin.

The Thévenin rules

The following three rules are used to determine the equivalent Thévenin circuit. We only have to determine the Thévenin voltage, V_{Th}, and the Thévenin resistance R_{Th}. The three rules are illustrated in Fig. 4.6.

Thévenin rule 1 (used to find V_{Th})

We have no load across the output terminals a and b, i.e., the green box is not connected to anything. We have an "*open circuit*". Hence, we do not have current flowing through R_{Th} and no potential drop through R_{Th}.

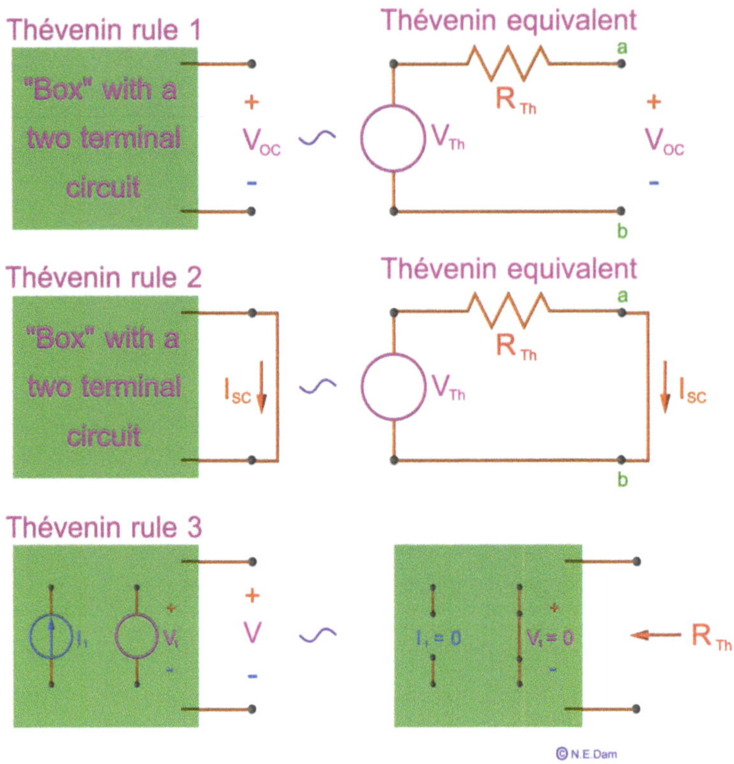

Fig. 4.6 The Thévenin rules used to determine V_{Th} and R_{Th} in the Thévenin equivalent circuit

When we measure or calculate the open circuit output voltage V_{OC}*, it is equal to the Thévenin voltage* V_{Th}.

$$V_{Th} = V_{OC} \tag{4.26}$$

Thévenin rule 2 (used to find R_{Th})

We now have a *"short circuit"* at the output terminals a and b, i.e., the green box is short circuited. We measure or calculate the short circuit current, I_{SC}, that flows from a to b between the output terminals. The potential drop across R_{Th} is V_{Th}. The current through R_{Th} is I_{SC}. Since we already know V_{Th}, we may determine R_{Th} using the Ohms law for the resistance R_{Th}.

When we measure or calculate the short circuit output current, I_{SC}*, it can be used to find the Thévenin resistance* R_{Th} .

$$R_{Th} = \frac{V_{Th}}{I_{SC}} \tag{4.27}$$

Thévenin rule 3 (used to find R_{Th})

We "zero" all independent voltage and current generators inside the green box.

Zeroing a voltage generator means that the voltage generator is replaced by a short circuit as illustrated in Fig. 4.6.

Zeroing a current generator, the current generator is replaced by an open circuit as illustrated in Fig. 4.6.

We zero all independent voltage and current generators inside the green box. Then we measure or calculate the resistance R_{Th} , which we "see" when we look into the output terminals.

EXAMPLE SHOWING HOW THE THÉVENIN RULES ARE USED

Thévenin rule 1

We use Thévenin rule 1 and calculate (or measure) the output voltage when we have no load across the output terminals, see the open circuit in the upper part of Fig. 4.7. We use the voltage divider formula to calculate the potential drop across R_2 and find

$$V_{Th} = V_{OC} = V_S \frac{R_2}{R_1 + R_2} \tag{4.28}$$

Thévenin rule 2

We use Thévenin rule 2 and calculate (or measure) the output current when we have a short circuit between the output terminals; see the short circuit in the lower part of Fig. 4.7. We may disregard R_2 when we have the short circuit because all the current flows through the short circuit and nothing through R_2, i.e., $I_2 = 0$. First, we calculate the short circuit current

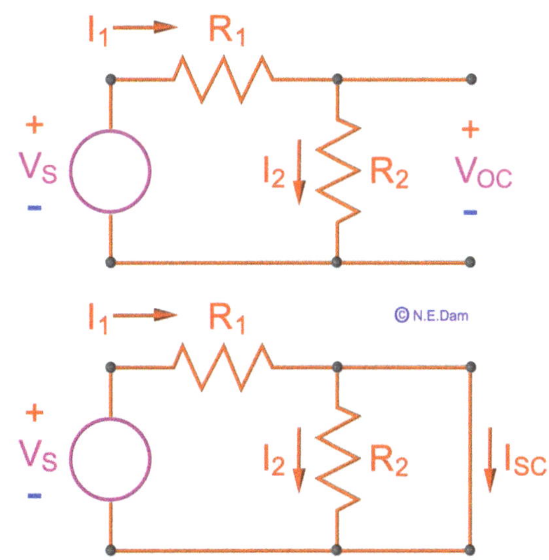

Fig. 4.7 Example showing how to use the Thévenin rules

$$I_{SC} = \frac{V_S}{R_1} \quad (4.29)$$

which means

$$R_{Th} = \frac{V_{Th}}{I_{SC}} = V_{Th}\frac{R_1}{V_S} \quad (4.30)$$

If we substitute V_{Th} from Eq. 4.28 into Eq. 4.30, we end up with

$$R_{Th} = V_{Th}\frac{R_1}{V_S} = V_S\frac{R_2}{R_1 + R_2}\frac{R_1}{V_S} = \frac{R_1 R_2}{R_1 + R_2} \quad (4.31)$$

Thévenin rule 3
Thévenin rule 3 uses zeroing of sources. There is only one source in this circuit. We zero the voltage source V_S by replacing the source with a short circuit. Thus, R_1 becomes parallel to R_2 in the upper part of Fig. 4.7. If we examine the two terminals on the right side of the circuit, we can see or measure the following resistance

$$R_{Th} = \frac{R_1 R_2}{R_1 + R_2} \quad (4.32)$$

This ends the treatment of the Thévenin equivalent circuit.

4.7 Norton Equivalent Circuit

The green box in Fig. 4.8 is considered a two terminal circuit with resistances together with voltage and/or current generators. If there are dependent generators in the circuit, they should depend on voltages and currents inside the circuit, i.e. inside the green box.

The two terminal circuit in the green box may be replaced by a simple equivalent circuit, the Norton equivalent circuit. If we look at the green box from outside the green box, the equivalent circuit has the same properties as the green box, i.e. it has the same voltage V across the output terminals a and b, and the current out of the terminals will be the same as for the green box.

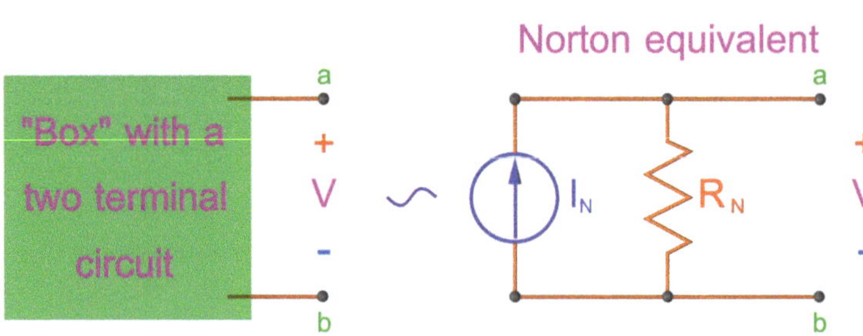

Fig. 4.8 The Norton equivalent circuit. N.E.Dam ©

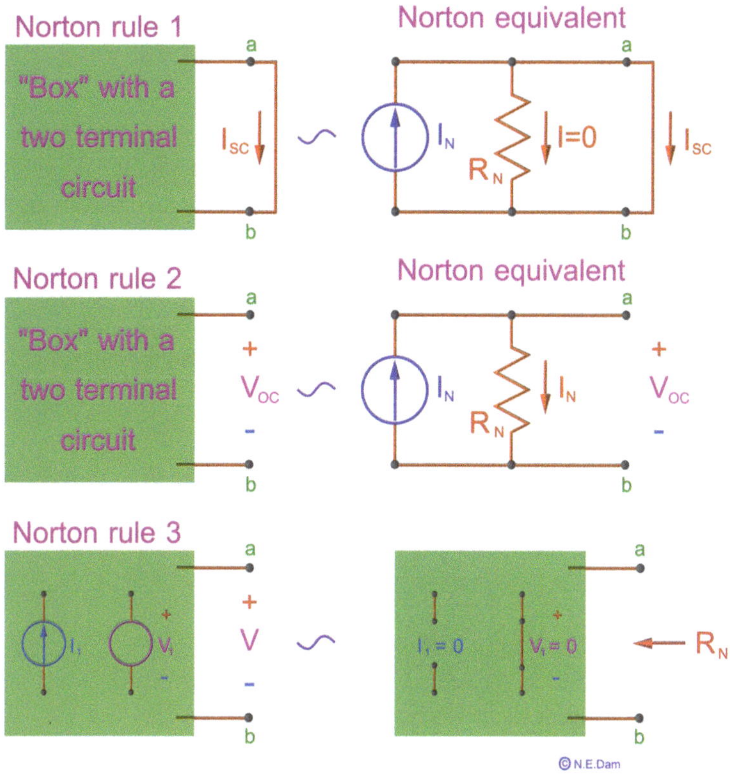

Fig. 4.9 The Norton rules used to determine I_N and G_N in the Norton equivalent circuit

The Norton equivalent circuit is an alternative to the Thévenin equivalent circuit, and the purpose of using the Norton equivalent circuit is the same as that of the Thévenin equivalent circuit. Sometimes it is most convenient to use the Norton equivalent circuit, and sometimes it is the Thévenin equivalent circuit that is most convenient to use. The Norton equivalent circuit was published by Edward Lawry Norton in 1926.

The Norton rules

The following three rules are used to determine the equivalent Norton circuit. We only have to determine the Norton current I_N. Norton resistance R_N turns out to be equal to Thévenin resistance R_{Th}. The three rules are illustrated in Fig. 4.9.

Norton rule 1 (used to find I_N)

We start with a short circuit across the output terminals a and b, i.e. the green box is short circuited. We measure or calculate the short circuit current, I_{SC} , that flows from a to b between the output terminals. When we have a short circuit on R_N, we may neglect the current through R_N. From the figure, we now see that $I_N = I_{SC}$.

When we measure or calculate the short circuit current, it equals the Norton current

$$I_N = I_{SC} \tag{4.33}$$

Norton rule 2 (used to find $R_N = R_{Th}$)
Now, we remove the short circuit and then we have no load across the output terminals a and b, i.e., the green box is not connected to anything. We have an open circuit. Hence, the current through R_N must be equal to I_N and the output voltage V_{OC} must be equal to the potential drop through R_N.

$$V_{OC} = R_N I_N = R_N I_{SC} \tag{4.34}$$

If we combine this equation by Eq. 4.26 and Eq. 4.27, we find the following.

$$V_{OC} = R_{Th} I_{SC} = R_N I_{SC} \Rightarrow R_{Th} = R_N \tag{4.35}$$

When we measure or calculate the open circuit output voltage V_{OC}, *it can be used to find the Norton resistance* R_N

$$R_N = \frac{V_{OC}}{I_{SC}} \tag{4.36}$$

Norton rule 3 (alternative way to find R_N)
This rule is exactly the same as Thévenin Rule 3. We zero all independent voltage generators and current generators inside the green box.
When we zero a voltage generator, the voltage generator is replaced by a short circuit as illustrated in Fig. 4.9.
When we zero a current generator, the current generator is replaced by an open circuit as illustrated in Fig. 4.9.

We zero all independent voltage generators and current generators inside the green box, then measure or calculate the resistance R_{Th} *that we "see" when we "look" at the output terminals from outside.*

4.8 Generator Transformation

Transformation from the Thévenin to the Norton equivalent circuit
If we know the Thévenin equivalent circuit, we may wish to find the Norton equivalent circuit if the Norton equivalent circuit is more convenient to use in the actual situation. We have already shown that

$$\boldsymbol{R_N = R_{Th}} \tag{4.37}$$

Figure 4.10 shows the transformation from the Thévenin circuit to the Norton equivalent. The figure may also be used in the reverse direction to transform from the Norton to the Thévenin equivalent circuit. We will return to that later.

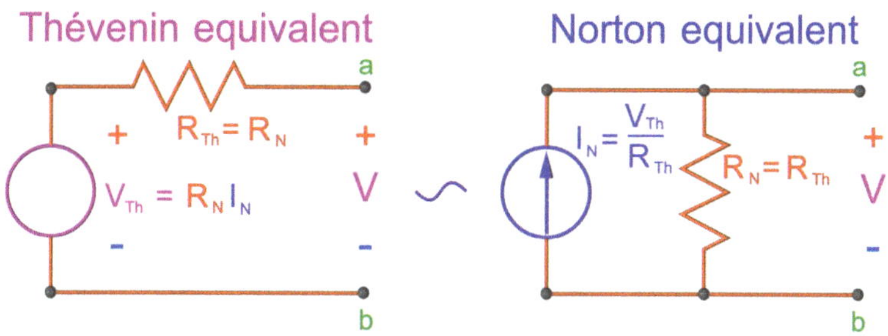

Fig. 4.10 Transformation from Thévenin to Norton equivalent circuit. N.E.Dam ©

First, we consider the middle part of Fig. 4.6. We can express the short-circuit current as follows.

$$I_{SC} = \frac{V_{Th}}{R_{Th}}$$

Now, we consider the upper part of Fig. 4.9. Here, we have a short circuit of the same green box as in Fig. 4.6. We see from the upper part of Fig. 4.9 that this short circuit current I_{SC} equals the Norton current I_N. Hence, we have

$$\boldsymbol{I_N = \frac{V_{Th}}{R_{Th}}} \tag{4.38}$$

Transformation from the Norton to the Thévenin equivalent circuit

From Eq. 4.37 we find

$$\boldsymbol{R_{Th} = R_N} \tag{4.39}$$

From Eq. 4.38 we find

$$\boldsymbol{V_{Th} = R_{Th} I_N = R_N I_N} \tag{4.40}$$

Remarks

After the transformation, we have the same output current and the same output voltage across the output terminals for the Thévenin and Norton equivalent circuits. However, the currents through R_{Th} and R_N are not the same for the two equivalent circuits.

For the Thévenin circuit in Fig. 4.10, the current through R_{Th} is identical to the output current, which is not the case for the Norton circuit. Performing a transformation, we should maintain the correct polarity for V_{Th} and the correct reference direction for I_N. E.g., if the + sign for V is at the "a" marked terminal, the arrow for I_N should point towards the "a" marked terminal.

Generator transformations may greatly simplify many circuits. The following example shows how to use repeated generator transformations to simplify a circuit.

Example with generator transformations

The following example shows how generator transformations can be used to simplify a circuit. We therefore want the circuit (shown at the top of Fig. 4.11) to be simplified to a simple equivalent Thevenin circuit, which is shown at the bottom of the same figure. The procedure is as follows:

The question is then where to start. Maybe we are thinking of considering I_{S1} and R_2 as a Norton equivalent circuit and transforming this into a Thevenin circuit. But this does not work, as I_{S1} and R_2 are connected to both the left and right sides. If V_{S1} and R_1 did not exist, then we could consider I_{S1} and R_2 as Norton equivalent circuits and transform this into a Thevenin circuit.

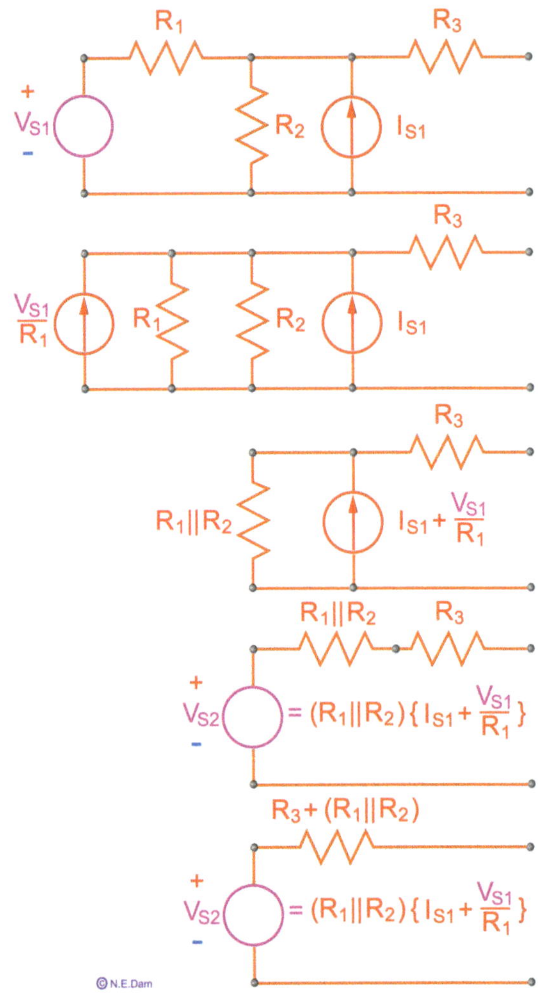

Fig. 4.11 Example showing how to simplify a circuit to a Thévenin equivalent circuit

The right place to start is on the left side of the circuit, where we identify a Thévenin circuit with V_{S1} and R_1, which is only connected to the right side. This circuit is transformed to a Norton circuit, where $I_N = \frac{V_{S1}}{R_1}$ in parallel with R_1. The current direction of I_N is upward as for the previous Thévenin circuit.

Now, R_1 is in parallel with R_2 and we may calculate the value of this parallel connection. It is denoted $R_1 || R_2$.

The two current generators are both upward directed and may be added to one current generator with the current

$$I_{S1} + \frac{V_{S1}}{R_1}$$

This Norton current generator is transformed into a Thévenin circuit with the Thévenin voltage

$$(R_1 || R_2)(I_{S1} + \frac{V_{S1}}{R_1})$$

and Thévenin resistance $R_1 || R_2$. The Thévenin resistance $R_1 || R_2$ is now added to R_3 giving the final Thévenin equivalent.

Final remarks

Many simple circuits may be solved using the methods described in Sects. 4.2 to 4.8. When we simplify series and parallel connections, when we use voltage and current dividers or when we make transformations between Thévenin and Norton equivalent circuits, we probably end up with a much simpler circuit which will give us the desired information. This could be a current or voltage across the output terminals of the circuit considered.

When simplifying the circuit, we reduce the number of circuit elements. Thereby, we also lose a lot of detailed information about voltages and currents through the removed circuit elements. Very often we don't need this information because we are perhaps only interested in a current or a voltage across the output terminals of the considered circuit.

However, sometimes we do want more detailed information about a current or a potential drop across some of the circuit elements than we may get from the simplifications described above. In that case, we should use a more general method to analyse the circuit. It is a method where we do not remove any of the circuit elements. Using this "node voltage method", we end up with information on the electric potentials at all nodes, and from these values we can calculate all the currents through all circuit elements. This node voltage method is described in the next section.

4.9 The Node Voltage Method

The analysis of this method comes in two parts:

1. The theoretical analysis of the circuit in the present section uses the Node voltage method which ends up in the node voltage Eqs. 4.45.
2. The mathematical analysis, where the starting point is the Node voltage Eqs. 4.45, takes place in Sect. 2.7 on page 35 followed by a practical example in Sect. 2.8 on page 37.

We start with a definition of a node:

A node is a point where two or more circuit elements meet.

With the node voltage method, the voltage (the electrical potential) is determined at each node. We are going to use the node voltage method to analyse the circuit in Fig. 4.12.

We identify 4 nodes, where 3 circuit elements meet.

Node 1 and 4

Node 4 is selected to be the reference node, where we define the potential to be equal to zero. This node is marked by the ground symbol. Often, we prefer the reference node to be at the minus terminal of the voltage generator. It is also an advantage to choose the reference node to be the one where as many circuit elements as possible meet. The voltage across R_4 is the potential at node 2 minus the potential at node 4

$$V_2 - V_4 = V_2 \text{ since } V_4 \equiv 0$$

Likewise for nodes 1 and 3. The unknown quantities that we want to find are V_1, V_2, V_3, i.e., all the node potentials. We find V_1 easily by observing $V_1 = V_{Gen}$. The remaining V_2 and V_3 are found by writing two independent equations using KCL. This will be elaborated on later.

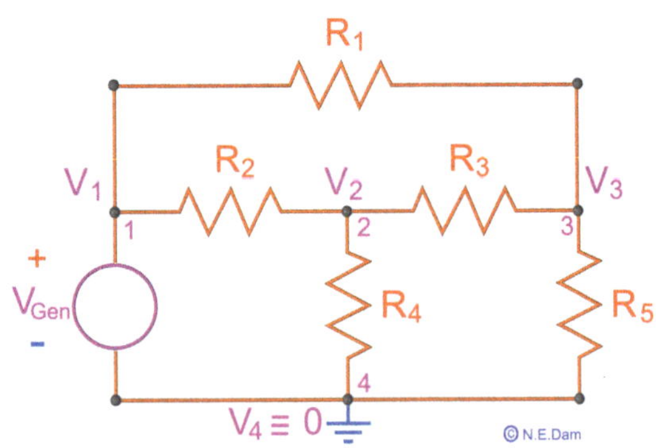

Fig. 4.12 Analyzing a circuit using the node voltage method

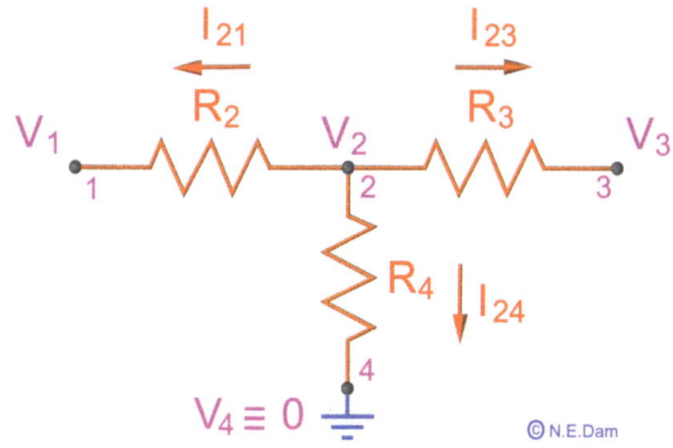

Fig. 4.13 Circuit elements and currents emanating from node 2

KCL for node 2

We consider the currents flowing away from node 2, see Fig. 4.13.

I_{21} flows from node 2 towards node 1 through R_2. I_{23} flows from node 2 towards node 3 through R_3. I_{24} flows from node 2 towards node 4 through R_4.

These three currents are considered positive if the positive charge flows away from node 2 (in the directions of the arrows). If, after completing the calculation, some of these currents end up being negative, then it is because these currents flow towards node 2. We have a passive configuration for R_2, R_3, and R_4. Since all three currents are orientated with arrow directions away from node 2, KCL for node 2 becomes:

$$I_{21} + I_{23} + I_{24} = 0 \tag{4.41}$$

Using Ohm's law for R_2, R_3, and R_4 we find:

$$I_{21} = \frac{V_2 - V_1}{R_2}, \quad I_{23} = \frac{V_2 - V_3}{R_3}, \quad I_{24} = \frac{V_2 - V_4}{R_4} \tag{4.42}$$

These currents are substituted into Eq. 4.41 and we find

$$I_{21} + I_{23} + I_{24} = \frac{V_2 - V_1}{R_2} + \frac{V_2 - V_3}{R_3} + \frac{V_2 - V_4}{R_4} = 0 \tag{4.43}$$

The described method for node 2 is thus:

- All currents are orientated with arrow directions away from node 2.
- For each current, we use Ohm's law.
- The sum of all the currents away from node 2 should be equal to zero according to KCL.

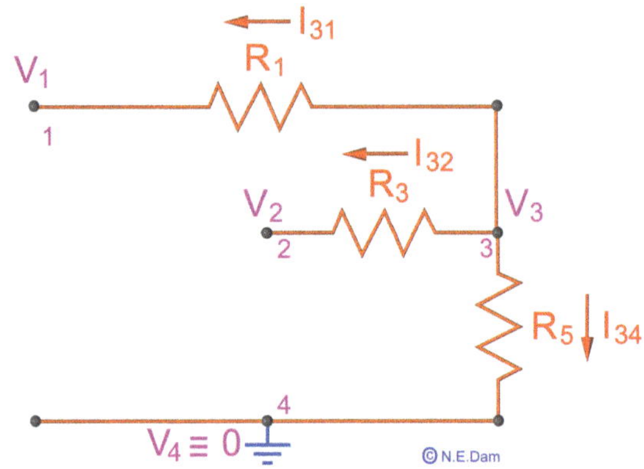

Fig. 4.14 Circuit elements and currents emanating from node 3

KCL for node 3

The method described above for node 2 can also be used for node 3 as shown in Fig. 4.14, so we find that

$$I_{31} + I_{32} + I_{34} = \frac{V_3 - V_1}{R_1} + \frac{V_3 - V_2}{R_3} + \frac{V_3 - V_4}{R_5} = 0 \tag{4.44}$$

The node voltage equations

We have set two equations (4.43 and 4.44) with two unknowns corresponding to the fact that we want to determine the potentials V_2 and V_3 for nodes 2 and 3:

$$\begin{aligned} \frac{V_2 - V_1}{R_2} + \frac{V_2 - V_3}{R_3} + \frac{V_2 - V_4}{R_4} = 0 \\ \frac{V_3 - V_1}{R_1} + \frac{V_3 - V_2}{R_3} + \frac{V_3 - V_4}{R_5} = 0 \\ V_1 = V_{Gen} \end{aligned} \tag{4.45}$$

We have hereby determined the two node voltage equations corresponding to the first part of the analysis (the circuit theoretical analysis) as described at the beginning of this section. The second part of the analysis (the mathematical analysis) starts with the two node voltage Eqs. 4.45 and ends up determining the two node potentials V_2 and V_3. This is done in Sect. 2.8 that illustrates how to use Cramer's method with two unknowns.

4.10 The Supernode Method

The supernode method is illustrated in Fig. 4.15. If we have a voltage generator between two nodes, we have a supernode. Nodes 1 and 2 and the V_{G1} voltage generator are one supernode. Nodes 3 and 4 and the V_{G2} voltage generator are another supernode.

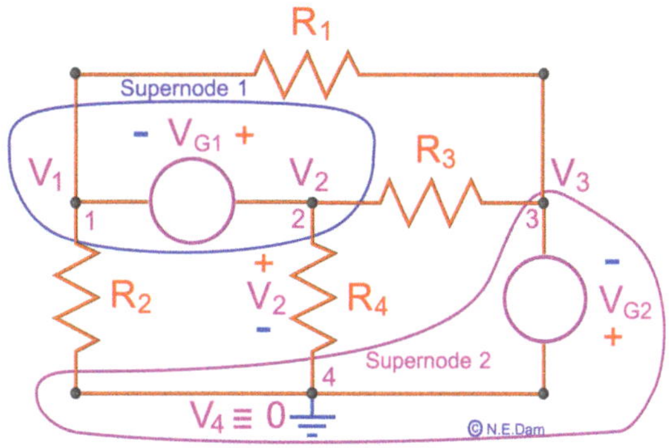

Fig. 4.15 Solving a circuit using the supernode method

This is how the supernode works between nodes 1 and 2. We use KCL starting from node 1.

1. The current from node 1 through R_1 is $\frac{V_1-V_3}{R_1}$.
2. The current from node 1 through R_2 is $\frac{V_1}{R_2}$.
3. The current from node 1 through V_{G1} is equal to the sum of the two currents through R_3 and R_4, which is $\frac{V_2-V_3}{R_3}$, respectively $\frac{V_2}{R_4}$

The sum of all these currents leaving node 1 is thus:

$$\frac{V_1 - V_3}{R_1} + \frac{V_1}{R_2} + \frac{V_2 - V_3}{R_3} + \frac{V_2}{R_4} = 0 \tag{4.46}$$

The potential at node 3 is

$$V_3 = -V_{G2} \tag{4.47}$$

We need one more equation to determine V_1 and V_2. If we write KCL for the other supernode, we get a dependent equation and no solution. Instead, we use the connection between node 1 and node 2:

$$V_2 = V_1 + V_{G1} \tag{4.48}$$

Now we substitute V_2 from Eq. 4.48 and V_3 from Eq. 4.47 into Eq. 4.46 and find

$$V_1\left(\frac{1}{R_1} + \frac{1}{R_2} + \frac{1}{R_3} + \frac{1}{R_4}\right) + V_{G1}\left(\frac{1}{R_3} + \frac{1}{R_4}\right) + V_{G2}\left(\frac{1}{R_1} + \frac{1}{R_3}\right) = 0 \tag{4.49}$$

From this equation, it is possible to find V_1. Thereafter we use Eq. 4.48 and find V_2. We assume that we know the values of the resistors and voltage generators.

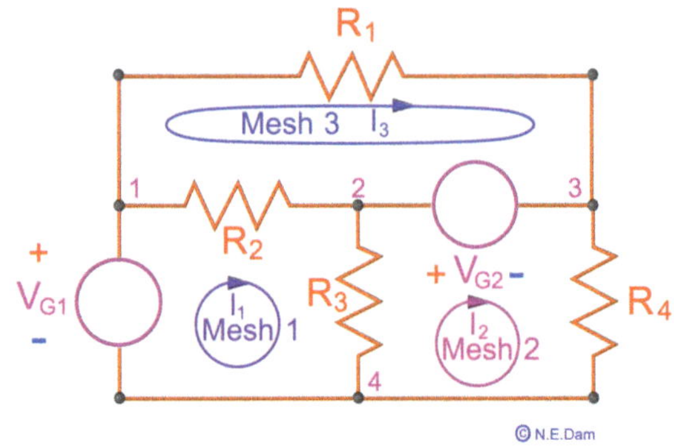

Fig. 4.16 Solving a circuit using the mesh-current method

4.11 The Mesh-Current Method

The analysis of this method comes in two parts:

1. The circuit theoretical analysis in the present Sect. 4.11, which ends up in the mesh current Eqs. 4.53.
2. The mathematical analysis, where the starting point is the mesh current Eqs. 4.53, takes place in Sect. 2.11 on page 44, which shows how to use Cramer's method with three unknowns. In Sect. 2.11, the mesh current Eqs. 4.53, now denoted 2.105, are reorganised into a matrix equation. The determinants are then calculated, and Cramer's method is used to find the final results for the mesh currents. In the example, we have three mesh currents which are determined using 3 by 3 determinants.

The mesh current method is illustrated with the example shown in Fig. 4.16. We identify three areas, where we introduce the so called mesh-currents $I_1,\ I_2,\ I_3$. We have a free choice of the mesh direction (as long as all are the same). In this example, we choose the clockwise direction.

Mesh 1
We "move" clockwise in mesh 1 using KVL, and the sum of potential increases equals the sum of potential decreases.

V_{G1} is a potential increase, since we pass through the voltage generator from minus to plus. Now, R_2 and R_3 are considered to give potential drops.
The potential drop across R_2 in the I_1 direction is $R_2(I_1 - I_3)$ since the net current through R_2 in the I_1 direction is $I_1 - I_3$. Similarly, the potential drop across R_3 in the I_1 direction is $R_3(I_1 - I_2)$. Finally, KVL becomes

$$V_{G1} = R_2(I_1 - I_3) + R_3(I_1 - I_2) \tag{4.50}$$

Mesh 2

We "move" clockwise in mesh 2 using KVL. V_{G2} is a potential drop, since we pass through the voltage generator from plus to minus. We find KVL:

$$V_{G2} + R_3(I_2 - I_1) + R_4 I_2 = 0 \tag{4.51}$$

Mesh 3

Finally, we "move" clockwise in mesh 3 using KVL. V_{G2} is a potential increase, since we pass through the voltage generator from minus to plus. We find KVL:

$$V_{G2} = R_2(I_3 - I_1) + R_1 I_3 \tag{4.52}$$

The mesh current equations

Rearranging Eqs. 4.50, 4.51 and 4.52 we find the *mesh current equations*:

$$\begin{aligned} (R_2 + R_3)I_1 \; - \; R_3 I_2 \; - \; R_2 I_3 &= V_{G1} \\ (-R_3)I_1 \; + \; (R_3 + R_4)I_2 \; + \; (0)I_3 &= -V_{G2} \\ (-R_2)I_1 \; + \; (0)I_2 \; + \; (R_1 + R_2)I_3 &= V_{G2} \end{aligned} \tag{4.53}$$

We have hereby determined the three mesh current equations corresponding to the first part of the analysis (the circuit theoretical analysis) as described at the beginning of this section. The second part of the analysis (the mathematical analysis) starts from the three mesh current equations and ends up determining the three mesh currents. This is done in Sect. 2.11 on page 44, which shows how to use Cramer's method with three unknowns.

4.12 The Δ Y Transformation

Motivation

Rather often, one encounters a resistive DC circuit, where a part of the entire circuit is a Δ-connected circuit (pronounced Delta-connected circuit), see Fig. 4.17. If the Δ-connected circuit is transformed to a Y-connected circuit it may be easier to simplify the entire circuit because the resistances in the Y-connected circuit can easily be combined with the surrounding resistances. This leads to a simplification of the entire circuit.

It can also be the other way around. If one encounters a resistive DC circuit, where a part of the circuit is a Y-connected circuit, it can be transformed into a Δ-connected circuit and perhaps simplify the entire circuit. In 3 phase AC circuits the load often consists of 3 resistances (or 3 impedances) connected as a Δ-connected circuit or as a Y-connected circuit, where the 3 phases are connected to the A, B, C terminals. In this case, the transformation is used to make an optimal connection to the generator.

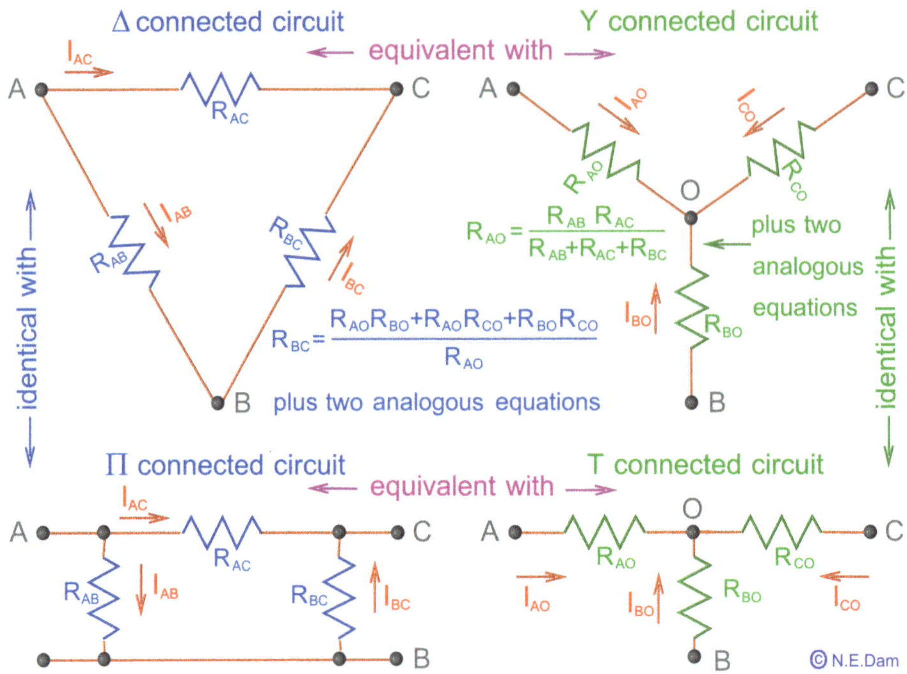

Fig. 4.17 A Δ connected circuit may be transformed into a Y-connected circuit and vice versa

Derivation of transformation formulas from a Δ to a Y-connected circuit

It is important that the transformation should leave all voltages and currents in the circuit outside the A-B-C triangle unchanged. If the circuit inside the A-B-C triangle is enclosed by a black box, we cannot detect if the circuit inside the black box is a Δ-connected circuit or a Y-connected circuit. All voltages and currents in the circuit outside the A-B-C triangle remain unchanged if we swap the circuit inside the black box from a Δ-connected circuit to a Y-connected circuit. This means that if we measure the resistance between the A and C terminals, it has the same value, no matter if we have a Δ-connected circuit or a Y-connected circuit.

If we measure or calculate the resistance between the A and C terminals in the Δ-connected circuit, we find the following resistance:

$$R_{AC\Delta} = R_{AC} \| (R_{AB} + R_{BC}) \tag{4.54}$$

The parallel connection between R_{AC} and $(R_{AB} + R_{BC})$ is denoted by $R_{AC\Delta}$.

If we measure or calculate the resistance between the A and B terminals in the Δ-connected circuit we find the following resistance:

$$R_{AB\Delta} = R_{AB} \| (R_{AC} + R_{BC}) \tag{4.55}$$

The parallel connection between R_{AB} and $(R_{AC} + R_{BC})$ is denoted by $R_{AB\Delta}$.

If we measure or calculate the resistance between the B and C terminals in the Δ-connected circuit we find the following resistance:

$$R_{BC\Delta} = R_{BC} \| (R_{AB} + R_{AC}) \tag{4.56}$$

The parallel connection between R_{BC} and $(R_{AB} + R_{AC})$ is denoted by $R_{BC\Delta}$.

If we measure or calculate the resistance between the A and C terminals in the Y-connected circuit we find the following resistance:

$$R_{ACY} = R_{AO} + R_{CO} \tag{4.57}$$

As mentioned above, if we measure the resistance between the A and C terminals, it has the same value, no matter if we have a Δ-connected circuit or a Y-connected circuit. It means that

$$R_{AC\Delta} = R_{ACY} = R_{AO} + R_{CO} \tag{4.58}$$

Likewise, we find for the resistance between the A and B terminals:

$$R_{AB\Delta} = R_{ABY} = R_{AO} + R_{BO} \tag{4.59}$$

Finally, we find for the resistance between the B and C terminals:

$$R_{BC\Delta} = R_{BCY} = R_{BO} + R_{CO} \tag{4.60}$$

In this section we consider the transformation from the Δ-connected circuit to the Y-connected circuit. We know R_{AB}, R_{AC}, R_{BC} and we want to find R_{AO}, R_{BO}, R_{CO}. Using the three equations, 4.58, 4.59, 4.60, we may write the following three equations

$$\begin{aligned}
1 \cdot R_{AO} + 0 \cdot R_{BO} + 1 \cdot R_{CO} &= R_{AC\Delta} \\
1 \cdot R_{AO} + 1 \cdot R_{BO} + 0 \cdot R_{CO} &= R_{AB\Delta} \\
0 \cdot R_{AO} + 1 \cdot R_{BO} + 1 \cdot R_{CO} &= R_{BC\Delta}
\end{aligned} \tag{4.61}$$

These three equations may be collected into the following matrix equation:

$$\begin{Bmatrix} 1\,0\,1 \\ 1\,1\,0 \\ 0\,1\,1 \end{Bmatrix} \begin{Bmatrix} R_{AO} \\ R_{BO} \\ R_{CO} \end{Bmatrix} = \begin{Bmatrix} R_{AC\Delta} \\ R_{AB\Delta} \\ R_{BC\Delta} \end{Bmatrix} \tag{4.62}$$

We find the three unknown R_{AO}, R_{BO}, R_{CO} by using Cramer's method. First, we calculate the determinant of the coefficient matrix:

$$A = \begin{vmatrix} 1 & 0 & 1 \\ 1 & 1 & 0 \\ 0 & 1 & 1 \end{vmatrix} = 2 \tag{4.63}$$

Then we may calculate

$$A_1 = \begin{vmatrix} R_{AC\Delta} & 0 & 1 \\ R_{AB\Delta} & 1 & 0 \\ R_{BC\Delta} & 1 & 1 \end{vmatrix}, \ A_2 = \begin{vmatrix} 1 & R_{AC\Delta} & 1 \\ 1 & R_{AB\Delta} & 0 \\ 0 & R_{BC\Delta} & 1 \end{vmatrix}, \ A_3 = \begin{vmatrix} 1 & 0 & R_{AC\Delta} \\ 1 & 1 & R_{AB\Delta} \\ 0 & 1 & R_{BC\Delta} \end{vmatrix} \tag{4.64}$$

According to Cramer's method, the solution is given by:

$$R_{AO} = \frac{A_1}{A}, \ R_{BO} = \frac{A_2}{A}, \ R_{CO} = \frac{A_3}{A} \tag{4.65}$$

We start with

$$A_1 = \begin{vmatrix} R_{AC\Delta} & 0 & 1 \\ R_{AB\Delta} & 1 & 0 \\ R_{BC\Delta} & 1 & 1 \end{vmatrix} = R_{AC\Delta} + R_{AB\Delta} - R_{BC\Delta} \tag{4.66}$$

We substitute 4.54, 4.55 and 4.56 into 4.66 and find

$$A_1 = R_{AC} \| (R_{AB} + R_{BC}) + R_{AB} \| (R_{AC} + R_{BC}) - R_{BC} \| (R_{AB} + R_{AC}) \tag{4.67}$$

This becomes

$$A_1 = \frac{R_{AC} \cdot (R_{AB} + R_{BC})}{R_{AC} + (R_{AB} + R_{BC})} + \frac{R_{AB} \cdot (R_{AC} + R_{BC})}{R_{AB} + (R_{AC} + R_{BC})} - \frac{R_{BC} \cdot (R_{AB} + R_{AC})}{R_{BC} + (R_{AB} + R_{AC})}$$

which reduces to

$$A_1 = \frac{2 \cdot R_{AB} \cdot R_{AC}}{R_{AB} + R_{AC} + R_{BC}} \tag{4.68}$$

Finally, we use the result from 4.63 and find

$$\boldsymbol{R_{AO} = \frac{A_1}{A} = \frac{R_{AB} \cdot R_{AC}}{R_{AB} + R_{AC} + R_{BC}}} \tag{4.69}$$

This formula is written in green in Fig. 4.17. If one should try to remember this result, one could use the following:

To find $\mathrm{R_{AO}}$ *in the Y-connected circuit, one should take the product of the two resistances connected to the A terminal in the* Δ*-connected circuit* $\mathrm{R_{AB} \cdot R_{AC}}$ *and divide by the sum of all three resistances in the* Δ*-connected circuit* $\mathrm{R_{AB} + R_{AC} + R_{BC}}$.

To find R_{BO} and R_{CO} one could calculate A_2 and A_3 in the same way as we did for A_1 and then use Eq. 4.65. However, it is not necessary to go through all these calculations again. It is much easier to use the result in Eq. 4.69 combined with the symmetry in the

Δ-connected circuit and in the Y-connected circuit. We find the following

$$R_{BO} = \frac{A_2}{A} = \frac{R_{AB} \cdot R_{BC}}{R_{AB} + R_{AC} + R_{BC}} \tag{4.70}$$

$$R_{CO} = \frac{A_3}{A} = \frac{R_{AC} \cdot R_{BC}}{R_{AB} + R_{AC} + R_{BC}} \tag{4.71}$$

Derivation of transformation formulas from a Y to a Δ-connected circuit

We use the results of Eqs. 4.69, 4.70, 4.71 and calculate the following sum:

$$R_{AO}R_{BO} + R_{AO}R_{CO} + R_{BO}R_{CO} = \frac{R_{AB}R_{AC}R_{AB}R_{BC} + R_{AB}R_{AC}R_{AC}R_{BC} + R_{AB}R_{BC}R_{AC}R_{BC}}{(R_{AB} + R_{AC} + R_{BC})^2} \tag{4.72}$$

which becomes

$$R_{AO}R_{BO} + R_{AO}R_{CO} + R_{BO}R_{CO} = \frac{R_{AB}R_{AC}R_{BC}(R_{AB} + R_{AC} + R_{BC})}{(R_{AB} + R_{AC} + R_{BC})^2} \tag{4.73}$$

and reduces further to

$$R_{AO}R_{BO} + R_{AO}R_{CO} + R_{BO}R_{CO} = \frac{R_{AB}R_{AC}R_{BC}}{R_{AB} + R_{AC} + R_{BC}} \tag{4.74}$$

We now observe that, according to Eq. 4.69, we have

$$R_{AO} = \frac{R_{AB} \cdot R_{AC}}{R_{AB} + R_{AC} + R_{BC}}$$

Thereby we find

$$R_{AO}R_{BO} + R_{AO}R_{CO} + R_{BO}R_{CO} = R_{AO}R_{BC} \tag{4.75}$$

and the final result becomes

$$\boldsymbol{R_{BC} = \frac{R_{AO}R_{BO} + R_{AO}R_{CO} + R_{BO}R_{CO}}{R_{AO}}} \tag{4.76}$$

This formula is written in blue in Fig. 4.17. If one should try to remember this result, one can use Fig. 4.17 and the following:

To find $\mathrm{R_{BC}}$ *in the Δ-connected circuit, one should take the sum of the products of the three resistances in the Y-connected circuit* $\mathrm{R_{AO}R_{BO} + R_{AO}R_{CO} + R_{BO}R_{CO}}$ *and divide by the resistance* $\mathrm{R_{AO}}$ *in the Y-connected circuit, which is "opposite" to* $\mathrm{R_{BC}}$ *in the Δ-connected circuit.*

Using symmetry, we find:

$$R_{AC} = \frac{R_{AO}R_{BO} + R_{AO}R_{CO} + R_{BO}R_{CO}}{R_{BO}} \tag{4.77}$$

and

$$R_{AB} = \frac{R_{AO}R_{BO} + R_{AO}R_{CO} + R_{BO}R_{CO}}{R_{CO}} \tag{4.78}$$

The Π-connected circuit

The Π-connected circuit in Fig. 4.17 is identical to the Δ-connected circuit. The 4 nodes at the bottom of the Π-connected circuit are, in fact, only one node (the B terminal).

The T-connected circuit

It is easy to see that the T-connected circuit in Fig. 4.17 is identical to the Y-connected circuit.

4.13 Superposition Method

Superposition is a method for simplifying the solution of a circuit with more independent and ideal generators.

Description of the method

We want to find the potential V_N of a specific node "N" or the current I_B through a specific branch "B". The principle is as follows:

1. We only consider one of the generators (number 1) and deactivate all other generators.[1] Then we simplify the circuit without removing the node "N" or without removing the branch "B" where we want to find the potential or the current. We determine the potential V_{N1} of the node "N" or the current I_{B1} through the branch "B".
2. Now we activate another generator (number 2) and deactivate all other generators. Again, we simplify the circuit and determine the potential V_{N2} of node "N" or the current I_{B2} through branch "B".
3. We repeat the process and activate all generators one by one while keeping all other generators deactivated. Each time we determine the potential of the node "N" or the current through the branch "B".

Finally, we find the potential V_N of a node "N":

$$V_N = V_{N1} + V_{N2} + \cdots \tag{4.79}$$

[1] Deactivation of independent ideal generators:
A voltage generator is deactivated by replacing it by a short circuit.
A current generator is deactivated by replacing it by an open circuit.
Dependent generators are never deactivated in the superposition method.

or we find the current I_B through branch "B":

$$I_B = I_{B1} + I_{B2} + \cdots \tag{4.80}$$

Please note that the superposition method does not always give a simpler solution, as sometimes it may result in even more equations to solve.

The superposition method can seem somewhat theoretical when, as above, it is described in general terms. The following example will show how the method is used in practice on a circuit.

After that, it is recommended to support understanding by solving a similar problem.

EXAMPLE

Figure 4.18 shows the original circuit. We are going to find V_2. Using the superposition method, we have the following two steps:

Step 1. The voltage generator V_{G1} is active, while V_{G2} is deactivated
We let V_{G1} be active while the voltage generator V_{G2} is deactivated, i.e., V_{G2} is replaced by a short circuit. Thus, we get Fig. 4.19, which is reduced to Fig. 4.20. Nodes 2 and 3 merge

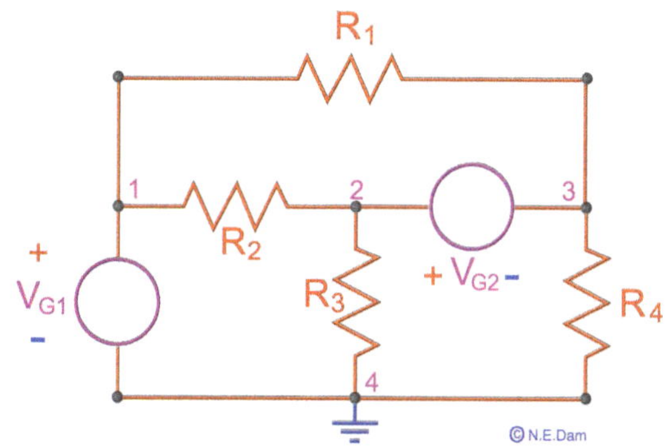

Fig. 4.18 The original circuit

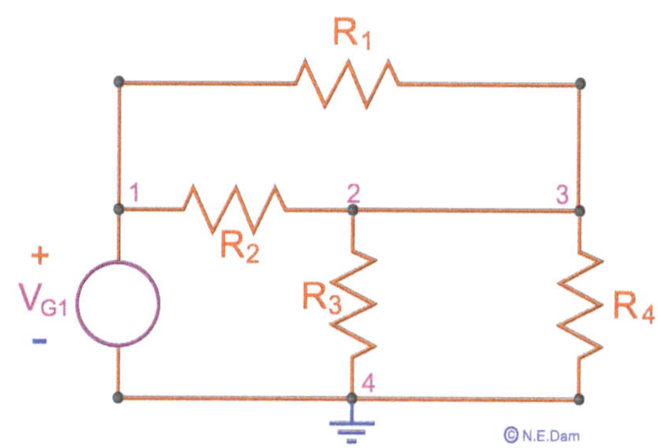

Fig. 4.19 Generator V_{G2} is deactivated

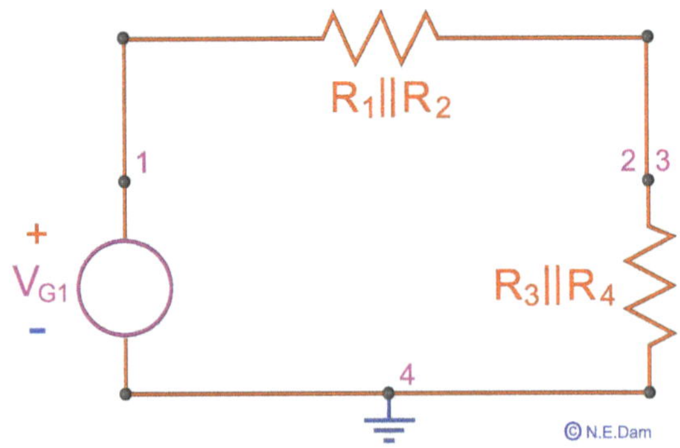

Fig. 4.20 Reduced version of Fig. 4.19

to one node, where we find the potential V_2 using the voltage divider formula:

$$V_{2,step1} = V_{G1}\frac{(R_3\|R_4)}{(R_1\|R_2) + (R_3\|R_4)} \tag{4.81}$$

Step 2. The voltage generator V_{G2} is active, while V_{G1} is deactivated

We let V_{G2} be active while the voltage generator V_{G1} is deactivated, i.e., V_{G1} is replaced by a short circuit. Hence, Fig. 4.18 is reduced to Fig. 4.21, which is reduced to Fig. 4.22. Nodes 1 and 4 merge into one node. Then we may reduce further to Fig. 4.23 because R_1 and R_4 are connected parallel.

To determine the potential at node 2, we temporarily ignore the ground connection and find the voltage across $(R_2\|R_3)$ using the voltage divider formula:

$$V_{(R_2\|R_3)} = V_{G2}\frac{(R_2\|R_3)}{(R_2\|R_3) + (R_1\|R_4)}$$

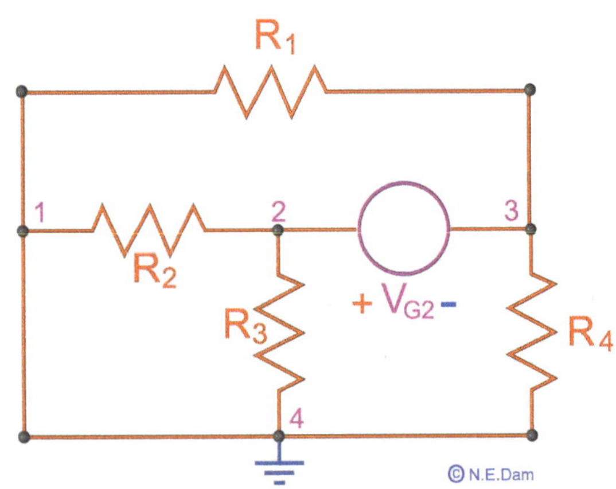

Fig. 4.21 Generator V_{G1} is deactivated

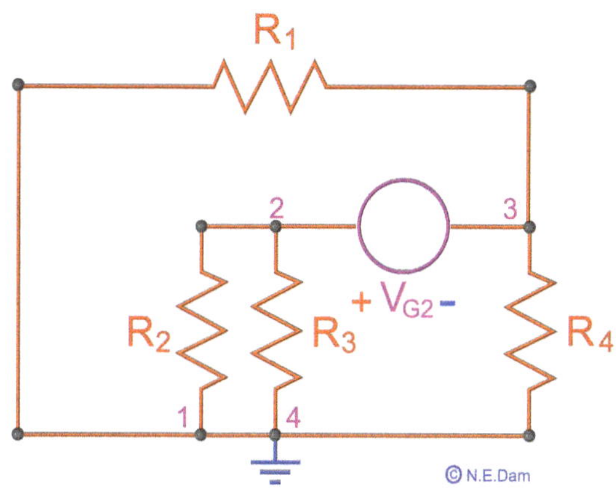

Fig. 4.22 Reduced version of Fig. 4.21

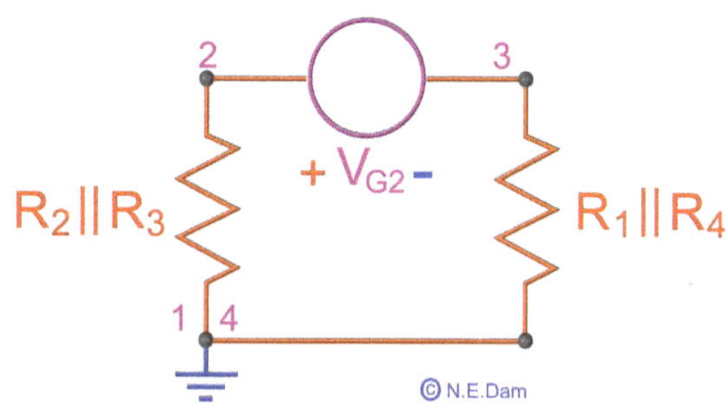

Fig. 4.23 Reduced version of Fig. 4.22

When we reconnect the ground to node 4, the entire circuit gets a reference potential. It is evident that the potential at node 2 is higher than the zero potential at node 4 due to the current direction in the circuit. The potential at node 2 must therefore be equal to the potential drop across $(R_2\|R_3)$:

$$V_{2,step2} = V_{(R_2\|R_3)} = V_{G2}\frac{(R_2\|R_3)}{(R_2\|R_3) + (R_1\|R_4)} \tag{4.82}$$

The final value for the potential at node 2 is therefore the sum of 4.81 and 4.82:

$$V_2 = V_{2,step1} + V_{2,step2} = V_{G1}\frac{(R_3\|R_4)}{(R_1\|R_2) + (R_3\|R_4)} + V_{G2}\frac{(R_2\|R_3)}{(R_2\|R_3) + (R_1\|R_4)} \tag{4.83}$$

4.14 Measurement of Resistance

There are several different methods for measuring resistance. In the following section, the Wheatstone bridge, 2-wire, 4-wire, and 3-wire sensing methods are described.

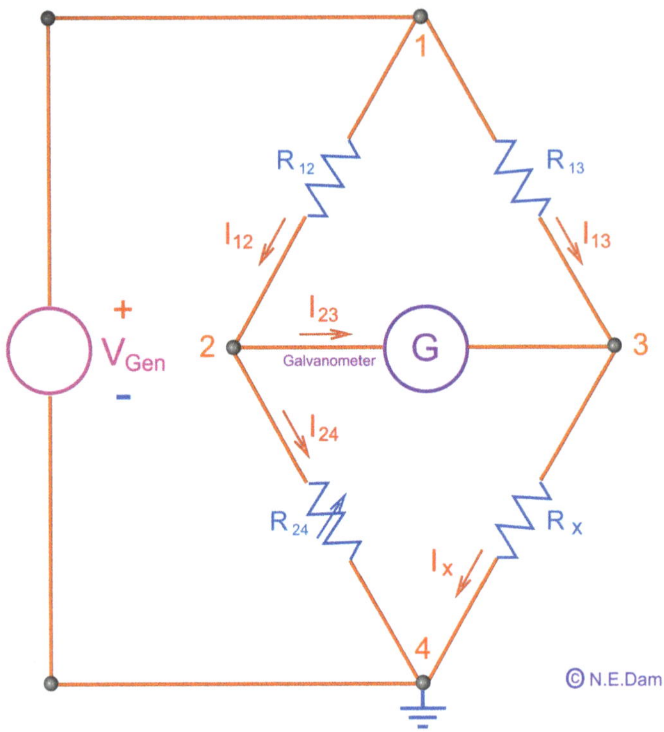

Fig. 4.24 The Wheatstone Bridge

4.14.1 The Wheatstone Bridge Method

This method is one of the most accurate methods for measuring resistance. Figure 4.24 shows the bridge circuit. This circuit operates with DC. However, it can be modified to operate with AC and impedances. The voltage generator supplies the circuit with the DC voltage V_{Gen}. The instrument marked with "G" is a galvanometer. It can measure very weak currents. One could also use a voltmeter. It should be able to detect a potential difference between nodes 2 and 3.

Adjustment of the Bridge

Adjusting the variable resistance R_{24} can bring the bridge in balance. The potential difference between nodes 2 and 3 is $V_{23} = V_2 - V_3$. The bridge may be very out of balance, i.e., V_{23} is relatively large. Here, it may be appropriate to use a voltmeter, which is set to show the large potential difference. We then adjust R_{24} so that we approach balance and V_{23} becomes smaller and smaller. The measuring range of the voltmeter is adjusted downward so that it can clearly show the smaller potential difference. When the bridge is in balance as possible, we can replace the voltmeter with a galvanometer that can show small currents. The last fine adjustment is then made so that the current in the bridge from node 2 to node 3 approaches zero, i.e., $I_{23} \to 0$.

When the Bridge is in Balance

The condition for the bridge to be in balance is

$$V_2 = V_3 \Rightarrow V_{23} = V_2 - V_3 = 0 \Rightarrow I_{23} = 0 \tag{4.84}$$

We define the potential of node 4 to be zero (ground). Then the potential at node 2 equals the potential drop from node 2 to node 4

$$V_2 = R_{24} I_{24} \tag{4.85}$$

Similarly, the potential at node 3 equals the potential drop from node 3 to node 4

$$V_3 = R_x I_x \tag{4.86}$$

When the bridge is in balance, we have $I_{23} = 0$. Using KCL we see

$$I_{12} = I_{24} \text{ and } I_{13} = I_x \tag{4.87}$$

If we set Eq. 4.85 equal to Eq. 4.86 and use Eq. 4.87 we find the following equation.

$$V_2 = V_3 \Rightarrow R_{24} I_{24} = R_x I_x \Rightarrow R_{24} I_{12} = R_x I_{13} \tag{4.88}$$

The condition $V_2 = V_3$ can also be expressed by the potential drop across R_{12} being equal to the potential drop across R_{13}

$$V_2 = V_3 \Rightarrow R_{12} I_{12} = R_{13} I_{13} \tag{4.89}$$

Now, we divide Eq. 4.88 by Eq. 4.89 and find

$$\frac{R_{24} I_{12}}{R_{12} I_{12}} = \frac{R_x I_{13}}{R_{13} I_{13}} \Rightarrow \frac{R_{24}}{R_{12}} = \frac{R_x}{R_{13}} \Rightarrow \tag{4.90}$$

$$\boldsymbol{R_x = R_{24} \frac{R_{13}}{R_{12}}} \tag{4.91}$$

If $R_{12} = R_{13}$ this result simplifies the equation

$$\boldsymbol{R_x = R_{24}} \tag{4.92}$$

Then the unknown resistor R_x is equal to what we read from the variable resistance R_{24}. This assumes that the variable resistor R_{24} is of the same order of magnitude as the unknown resistor R_x. That is, R_{24} can be adjusted to the same value as R_x. If, for example, R_x is about 10 times larger than R_{24}, one can choose R_{12} and R_{13} so that $\frac{R_{13}}{R_{12}} = 10$. Then R_{24} is multiplied by a factor 10 in Eq. 4.91.

Remarks

The variable resistor R_{24} can be a relatively accurate decade resistor. The total accuracy that can be achieved depends on the accuracy of all the resistors in the circuit. The accuracy of the measuring instrument is not significant as the instrument is simply used to show whether the bridge is in balance. In contrast, the sensitivity of the instrument is important. A very sensitive instrument will be able to show a very weak current in the bridge, which can be thereby adjusted to a more precise balance.

One must be aware that the heat development in the resistors of the circuit must be very small relative to what the resistors can withstand so that the resistance values do not change due to a change in temperature.

4.14.2 The 2-Wire Sensing Method

This method is the simplest (but not the most accurate) method of resistance measurement. Figure 4.25 shows the principle. A current generator and an ammeter are connected through two wires to the sample we want to measure. The two wires may be the red and blue (or

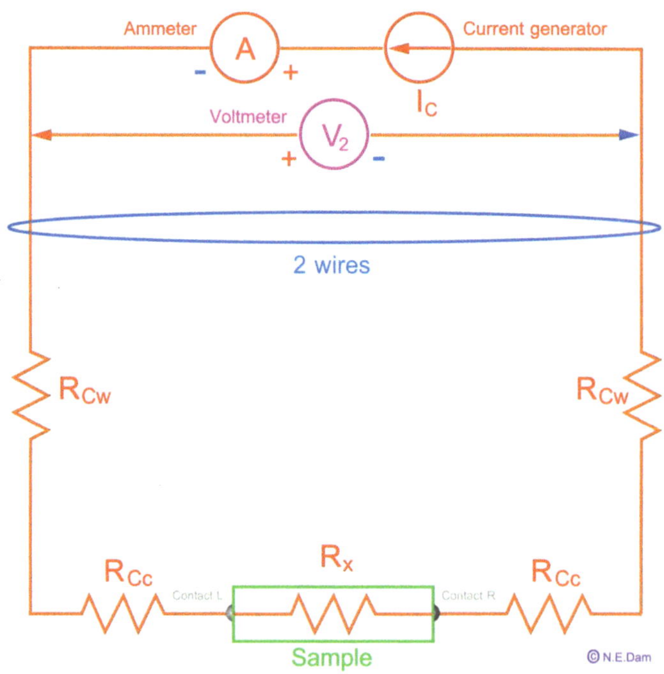

Fig. 4.25 The 2-wire sensing method

black) wires from an ohmmeter to the sample. Alternatively, the two wires may be enclosed in a long cable to the sample as indicated by the ellipse in Fig. 4.25.

The resistance of the wire from the measuring instrument to the sample is indicated by R_{Cw}, where C means the circuit with the **C**urrent generator and w means the **w**ire from the measuring instrument to the sample. The wire is connected to the sample. For example, the wire may be soldered to a metal sample. The transition between the wire and the sample is the soldering point. This contact has resistance R_{Cc}, where C means the circuit with the **C**urrent generator and c means the **c**ontact. This resistance is in the soldering point itself, but in the figure the resistance is shown outside the soldering point.

The measuring setup consists of a current generator, an ammeter, and a voltmeter which measures the potential difference V_2 between the two wires to the sample. This potential difference is also the potential drop across the resistors

$$R_{Cw} + R_{Cc} + R_x + R_{Cc} + R_{Cw}$$

where R_x is the resistance of the sample. If the current is adjusted to the value I_C then V_2 can be expressed by

$$V_2 = (2R_{Cw} + 2R_{Cc} + R_x)I_C \tag{4.93}$$

The total potential drop V_2 includes the potential drop across the sample resistance R_x, but also the potential drop across the two wires and the two contacts. If

$$2R_{Cw} + 2R_{Cc} \ll R_x \tag{4.94}$$

we may neglect the resistances of the two wires and the two contacts relative to the sample resistance R_x. This simplifies Eq. 4.93 to

$$V_2 = V_{R_x} = R_x I_C$$

and therefore

$$\boldsymbol{R_x = \frac{V_2}{I_C}} \tag{4.95}$$

For example, if $I_C = 1$ mA and we measure $V_2 = 1$ V, then $R_x = \frac{V_2}{I_C} = \frac{1\text{ V}}{1\text{ mA}} = 1\ \mathbf{k\Omega}$. Therefore, a reading of 1 V on the voltmeter corresponds to 1000 Ω. The voltmeter thus directly shows the resistance R_x of the sample if the reading of the voltmeter is multiplied by a factor 1000. If I_C is adjusted to other values, resistance can be measured in other ranges.

If the resistance of the wires and contacts is of the same order of magnitude as the resistance of the sample, we have the following.

$$2R_{Cw} + 2R_{Cc} \approx R_x \tag{4.96}$$

In this case, we cannot neglect the resistance of the two wires and the two contacts relative to the sample resistance R_x . If we carry out a measurement where we have shorted the sample, we have

$$V_{2,R_x=0} = (2R_{Cw} + 2R_{Cc})I_C$$

This can be used in Eq. 4.93. If we carry out another measurement where we have not shorted the sample, we find

$$V_2 = V_{2,R_x=0} + V_{R_x} = V_{2,R_x=0} + R_x I_C$$

Then we find

$$R_x I_C = V_2 - V_{2,R_x=0}$$

and thereby

$$\boldsymbol{R_x = \frac{V_2 - V_{2,R_x=0}}{I_C}} \tag{4.97}$$

When carrying out this two-step measurement, one can, with some approximation, eliminate the wire and contact resistance from the resistance measurement.

Resistance is often measured with an ohmmeter or a multimeter which can measure resistance. Here, there are typically only two wires from the measuring instrument to the sample. Therefore, we obtain the method described above in this section. Sometimes, the metre has an adjustment knob that is turned until the metre reads zero when the two leads to the sample are shorted (the two "probes" are shorted).

Please note that there may be some uncertainty in the determination of R_x if R_x is small relative to the resistance of the wires and contacts. Also note that all resistances depend on the temperature. This applies regardless of the measurement method. One should always note what temperatures the different parts of the circuit have when the measurement is carried out.

4.14.3 The 4-Wire Sensing Method

This is the most accurate method for measuring relatively small resistance values and is yet relatively simple to implement.

Figure 4.26 shows the principle of the method. We have in total 4 wires to the sample. The two outermost wires send the current I_C through the sample as with the 2-wire method. The voltmeter V_4 is connected to the two middle wires. In this circuit, we have a very weak current $I_V \ll I_C$. The voltmeter V_4 measures the potential drop across the sample from the left end surface to the right end surface.

The figure shows that the two middle wires are attached to the sample with the same soldering points as used in the current circuit (at the left and right ends, respectively). Alternatively, the two middle wires may be attached to the sample with separate soldering points located some distance from the left and right ends, respectively. The measured resistance R_x will in both cases be the resistance between the two soldering points that make up the two middle wires.

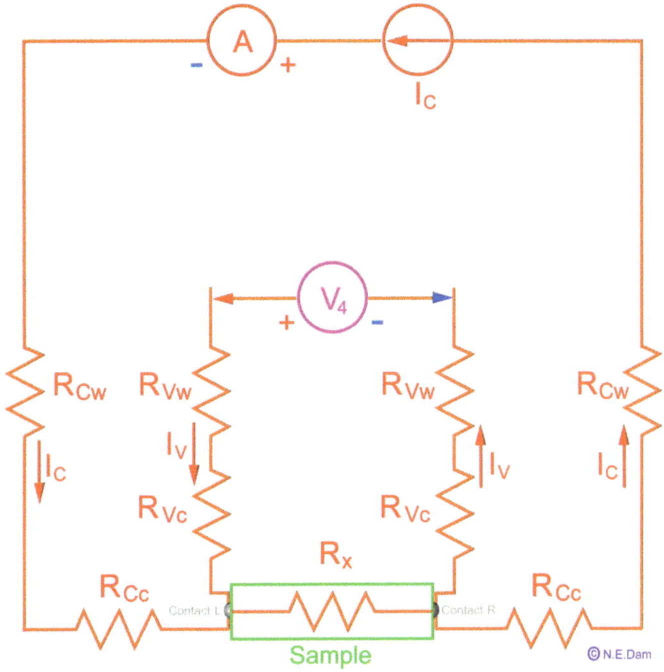

Fig. 4.26 The 4-wire sensing method

The voltmeter V_4 measures the following potential drop:

$$V_4 = (2R_{Vw} + 2R_{Vc})I_V + R_x(I_V + I_C) \tag{4.98}$$

The current I_V through the voltmeter V_4 is perhaps one million times less than the current I_C. It is therefore an extremely good approximation to disregard $(2R_{Vw} + 2R_{Vc})I_V$ and disregard I_V in $(I_V + I_C)$. We find

$$V_4 = (2R_{Vw} + 2R_{Vc})I_V + R_x(I_V + I_C) \cong R_x I_C \tag{4.99}$$

Thereby we find

$$R_x \cong \frac{V_4}{I_C} \tag{4.100}$$

This is an extremely good approximation and is the most accurate method to determine the unknown resistance R_x. It is also a relatively simple method. Measurements do not need to be made in several rounds, and no bridge should be brought into balance with variable resistance. The condition $I_V \ll I_C$ is easily met with a voltmeter that has a relatively high input resistance.

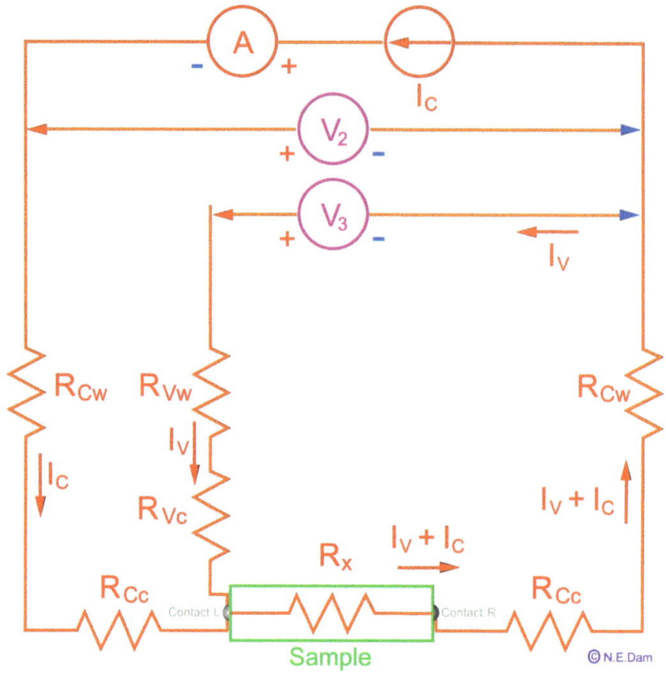

Fig. 4.27 The 3-wire sensing method

4.14.4 The 3-Wire Sensing Method

This method requires only 3 wires to the sample and is almost as accurate as the 4-wire sensing method for measuring relatively small resistance values. Some measuring instruments can perform the necessary calculations.

The reason we review the 3-wire sensing method after the 4-wire sensing method is that the 3-wire sensing method is more complicated than the 4-wire sensing method and is easier to understand once the 4-wire sensing method has been seen.

Figure 4.27 shows the principle of the method. We have in total 3 wires to the sample. The two outermost wires send the current I_C through the sample just as with the 2- and 4-wire methods. A very weak current $I_V \ll I_C$ flows through the voltmeter V_3 and the middle wire.

The voltmeter V_2 measures the following potential drop

$$V_2 = (R_{Cw} + R_{Cc})I_C + (R_x + R_{Cc} + R_{Cw})(I_V + I_C) \tag{4.101}$$

We neglect the terms with I_V and find

$$V_2 \cong (R_{Cw} + R_{Cc})I_C + (R_x + R_{Cc} + R_{Cw})I_C = (R_x + 2R_{Cc} + 2R_{Cw})I_C \tag{4.102}$$

The voltmeter V_3 measures the following potential drop

$$V_3 = (R_{Vw} + R_{Vc})I_V + (R_x + R_{Cc} + R_{Cw})(I_V + I_C) \tag{4.103}$$

We neglect the terms with I_V and find

$$V_3 \cong (R_x + R_{Cc} + R_{Cw})I_C \tag{4.104}$$

We form the difference

$$2V_3 - V_2 \cong (2R_x + 2R_{Cc} + 2R_{Cw})I_C - (R_x + 2R_{Cc} + 2R_{Cw})I_C = R_x I_C \tag{4.105}$$

Thereby we end up with

$$R_x \cong \frac{2V_3 - V_2}{I_C} \tag{4.106}$$

This requires a current measurement of I_C as well as two voltage measurements of V_2 and V_3, followed by the calculation in Eq. 4.106.

It is possible to buy a device with the necessary measuring instruments, as well as a microprocessor. This device is dedicated to the 3-wire sensing method. The advantage of the 3-wire sensing method compared to the 4-wire sensing method is, of course, that a wire in the cable between the device and R_x can be saved, which is important for longer cable runs.

4.15 Tests

General guidelines for answering the test

The test is without any aids; only pencils, rulers, etc. are permitted. The evaluation of the test, which is handed in, reflects not only the correctness of the answer but also the care and dedication that has been put into the answer.

A. Write the answer or formula correctly and show all the steps in the calculations.
B. The answer must be well structured, orderly, and easy to read.
C. Briefly explain the symbols used the first time they appear in the test.
D. If suggested, make a small sketch or diagram in colour to add clarity.

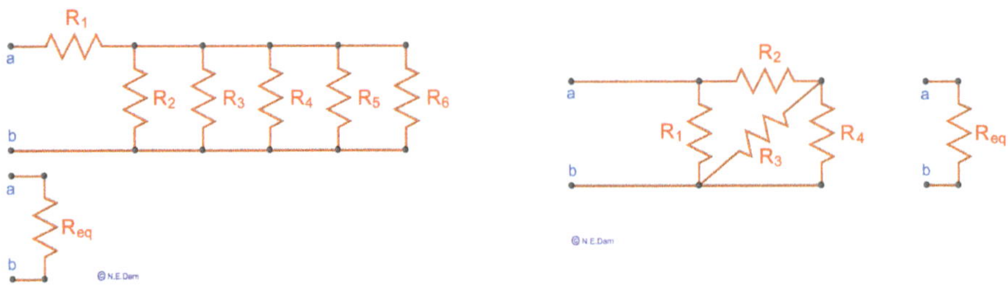

(a) Parallel connection of resistors.

(b) Simplification of a circuit

Fig. 4.28 Figures to be used in problems 1 and 2

4.15.1 Test 4.1 Basic Problems with Resistors

Problem 1 (20%) Parallel connection of resistors

For the circuit in Fig. 4.28a , we have $R_1 = R_2 = 1\ \Omega$, $R_3 = R_4 = R_5 = R_6 = 4\ \Omega$

A. Find R_{eq} and illustrate the steps towards the solution with sketches. Do not use conductances.
B. Find R_{eq} and illustrate the steps towards the solution with sketches. Use the conductances for R_2 to R_6 when calculating the parallel connection.

Problem 2 (20%) Simplification of a circuit

For the circuit in Fig. 4.28b , we have $R_1 = 5\ \Omega$, $R_2 = 3,5\ \Omega$, $R_3 = R_4 = 3\ \Omega$

A. Find R_{eq} and illustrate the steps towards the solution with sketches.

Problem 3 (30%) Thévenin equivalent circuit

For the circuit in Fig. 4.29a , we have $R_1 = 200\ \Omega$, $R_2 = 300\ \Omega$, $V_S = 10$ V

A. Use rule Th1 and find V_{Th} using V_{OC} and the voltage divider formula.
B. Use rule Th2 and find I_{SC} and R_{Th}, note that $I_2 = 0$.
C. Use rule Th3 and find R_{Th} by zeroing the sources.
D. Sketch the equivalent Thévenin circuit.

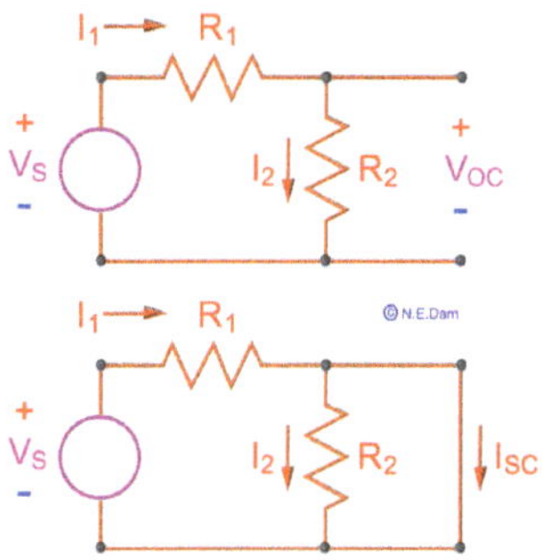

(a) Thévenin equivalent circuit.

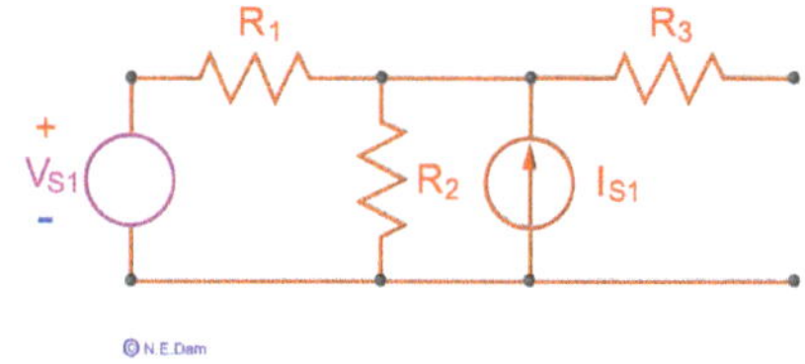

(b) Simplification of a circuit using transformations between Thévenin and Norton equivalent circuits.

Fig. 4.29 Figures to be used in problems 3 and 4

Problem 4 (30%) Simplification of a circuit using transformations between Thévenin and Norton equivalent circuits.

For the circuit in Fig. 4.29b , we have $R_1 = R_2 = 8\ \Omega$, $R_3 = 4\ \Omega$, $I_{S1} = 3$ A, $V_{S1} = 24$ V

A. Show all the steps in reducing the circuit to a final Thévenin equivalent circuit.

4.15.2 Test 4.2 The Node and Supernode Voltage Method

Problem 1 (50%) The Node voltage method

For the circuit in Fig. 4.30, we have $R_1 = R_2 = R_3 = R_4 = R_5 = 10\ \Omega$, $V_{Gen} = 10\ V$

Fig. 4.30 Circuit to be solved by the Node voltage method

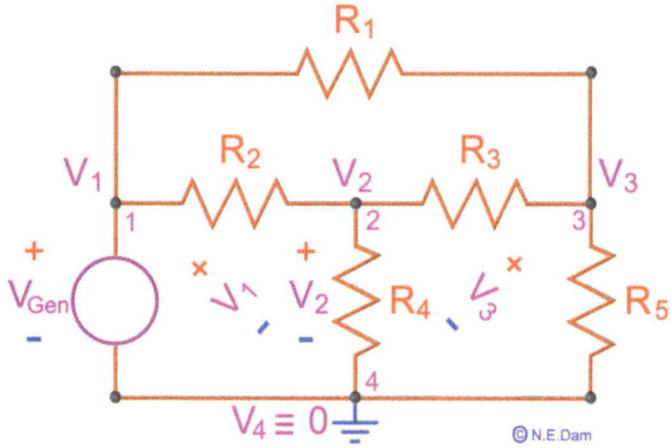

A. For the circuit, write the equations to find $V_1,\ V_2,\ V_3$ using the Node voltage method and explain the method. Use the symbols (no component values) and do not solve the equations, only write the two equations to find $V_2,$ and V_3.
B. Now use the listed values for the resistors and the voltage generator and solve the two equations to find $V_2,$ and V_3 by hand and show each step using the Cramer's method. Remember to state the value of V_1.

Problem 2 (50%) The Supernode voltage method

For the circuit in Fig. 4.31, we have $R_1 = R_2 = R_3 = R_4 = 10\ \Omega,\ V_{G1} = V_{G2} = 10\ \text{V}$

A. For the circuit, write the equations to find $V_1,\ V_2,\ V_3$ using the supernode voltage method and explain the method. You should use the symbols (no component values) and you should not solve the equations, only write the two equations to find $V_1,$ and V_2.
B. Now use the listed values for the resistors and voltage generators and solve the two equations to find $V_1,$ and V_2 by hand and show each step. Remember to state the value of V_3.

4.15.3 Test 4.3 The Mesh Current Method

Before answering the following questions, it is recommended to read the following

- Section 2.10 on page 41
- Section 2.11 on page 44
- Sect. 4.11 on page 108

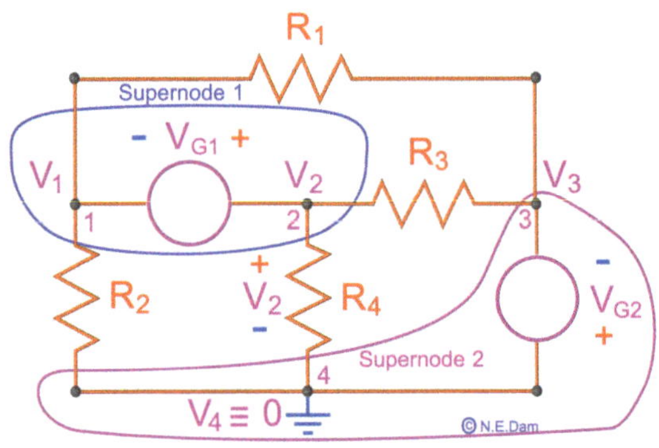

Fig. 4.31 Circuit to be solved by the Supernode method

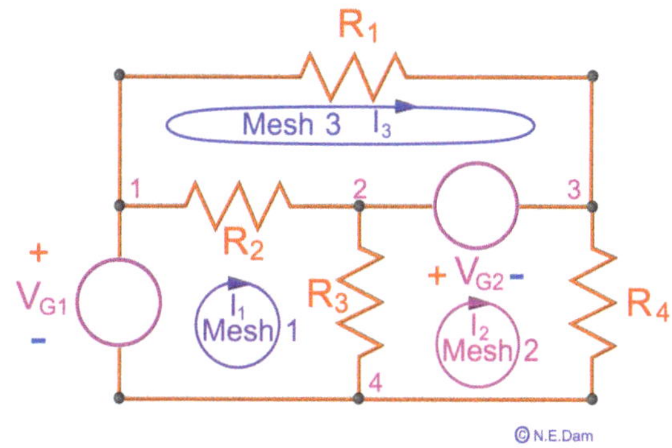

Fig. 4.32 Solving a circuit using mesh currents and Cramer's method

We define the three meshes in Fig. 4.32 with the three mesh currents $I_1,\ I_2,\ I_3$. These three currents are not identical to the real currents flowing through the resistances. For example, the real current flowing through R_2 from left to right is $I_1 - I_3$.

Problem 1 (8%) Kirchhoff's Voltage Law

Describe Kirchhoff's Voltage Law (KVL)

Problem 2 (8%) Kirchhoff's Voltage Law for mesh 1

Use the method in Sect. 2.11 on page 44 to prove that KVL for mesh 1 becomes
$V_{G1} = R_2(I_1 - I_3) + R_3(I_1 - I_2)$
Rearrange this equation and prove that it is equivalent with:
$(R_2 + R_3)I_1 \ - \ R_3I_2 \ - \ R_2I_3 = V_{G1}$

Problem 3 (8%) Kirchhoff's Voltage Law for mesh 2

Repeat the method for mesh 2.

Problem 4 (8%) Kirchhoff's Voltage Law for mesh 3

Repeat the method for mesh 3.

Problem 5 (4%) The rearranged KVL equations

Write the three rearranged KVL equations from problems 2–4 below each other.

Problem 6 (4%) The matrix equation

Write the three rearranged KVL equations as a matrix equation, which can be used later to find the three mesh-currents $I_1,\ I_2,\ I_3$.

In this example, we have the following

$$V_{G1} = V_{G2} = 20\ \text{V},\ R_1 = 2\ \Omega,\ R_2 = 4\ \Omega,\ R_3 = 6\ \Omega,\ R_4 = 8\ \Omega$$

It should be noted that if we multiply a matrix by a real or complex constant c, it has the effect that c multiplies on all elements of the matrix:

$$c\mathbf{A} = c\begin{Bmatrix} a_{11} & a_{12} & a_{13} \\ a_{21} & a_{22} & a_{23} \\ a_{31} & a_{32} & a_{33} \end{Bmatrix} = \begin{Bmatrix} ca_{11} & ca_{12} & ca_{13} \\ ca_{21} & ca_{22} & ca_{23} \\ ca_{31} & ca_{32} & ca_{33} \end{Bmatrix}$$

Problem 7 (12%) The matrix equation with the above figures

Write the matrix equation with the above figures and calculate the determinant A for the coefficient matrix.

It should be noted that if we multiply a determinant by a real or complex constant c, it has the effect that c multiplies on all elements in one row or all elements in one column:

$$cA = c\begin{vmatrix} a_{11} & a_{12} & a_{13} \\ a_{21} & a_{22} & a_{23} \\ a_{31} & a_{32} & a_{33} \end{vmatrix} = \begin{vmatrix} ca_{11} & ca_{12} & ca_{13} \\ a_{21} & a_{22} & a_{23} \\ a_{31} & a_{32} & a_{33} \end{vmatrix} = \begin{vmatrix} a_{11} & a_{12} & a_{13} \\ ca_{21} & ca_{22} & ca_{23} \\ a_{31} & a_{32} & a_{33} \end{vmatrix} = \begin{vmatrix} a_{11} & a_{12} & a_{13} \\ a_{21} & a_{22} & a_{23} \\ ca_{31} & ca_{32} & ca_{33} \end{vmatrix}$$

or

$$cA = c\begin{vmatrix} a_{11} & a_{12} & a_{13} \\ a_{21} & a_{22} & a_{23} \\ a_{31} & a_{32} & a_{33} \end{vmatrix} = \begin{vmatrix} ca_{11} & a_{12} & a_{13} \\ ca_{21} & a_{22} & a_{23} \\ ca_{31} & a_{32} & a_{33} \end{vmatrix} = \begin{vmatrix} a_{11} & ca_{12} & a_{13} \\ a_{21} & ca_{22} & a_{23} \\ a_{31} & ca_{32} & a_{33} \end{vmatrix} = \begin{vmatrix} a_{11} & a_{12} & ca_{13} \\ a_{21} & a_{22} & ca_{23} \\ a_{31} & a_{32} & ca_{33} \end{vmatrix}$$

Problem 8 (24%) Calculation of determinants

Calculate the three determinants $A_1,\ A_2,\ A_3$ that should be used in the Cramer method.

Problem 9 (12%) Calculation of mesh currents

Calculate the three mesh currents $I_1,\ I_2,\ I_3$

Problem 10 (4%) Calculation of the current through R_2

Calculate the current through R_2 towards the right and find the real current direction.

Problem 11 (8%) Calculation of the potential at node 2

Let the potential at node 4 be zero. Find the potential V_2 at node 2 relative to node 4. NB! One way to do this is to use the result from question 10 (Fig. 4.33).

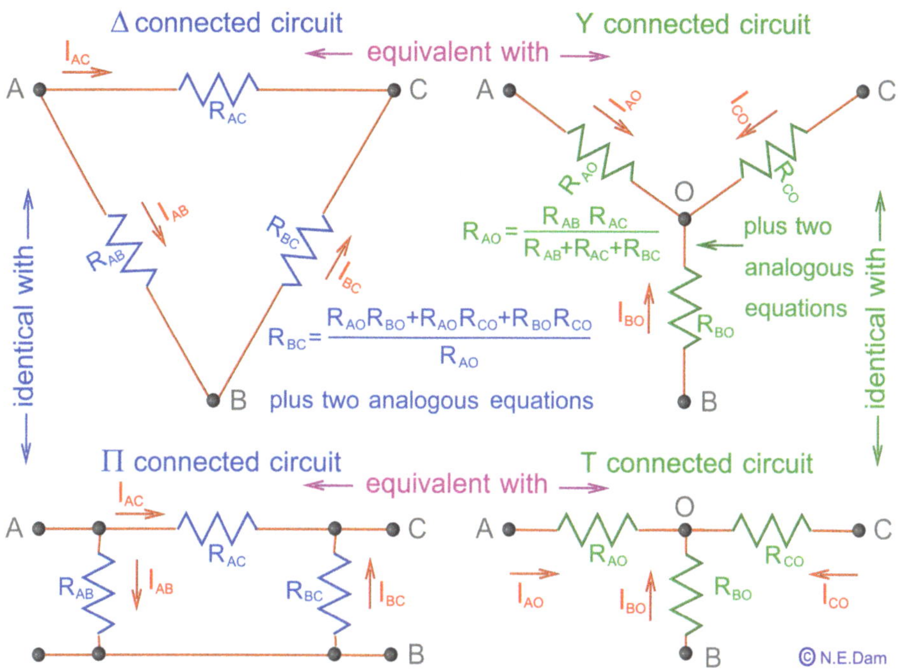

Fig. 4.33 A Δ-connected circuit may be transformed into a Y-connected circuit and reverse

4.15.4 Test 4.4 Δ Y and Y Δ Transformations

Problem 1 (40%) Formula for the Δ - Y transformation

Prove the following formula for the Δ - Y transformation using Cramer's method:

$$R_{AO} = \frac{R_{AB} \cdot R_{AC}}{R_{AB} + R_{AC} + R_{BC}} \tag{4.107}$$

Problem 2 (30%) Formula for the Y - Δ transformation

Prove the following formula for the Y - Δ transformation:

$$R_{BC} = \frac{R_{AO}R_{BO} + R_{AO}R_{CO} + R_{BO}R_{CO}}{R_{AO}} \tag{4.108}$$

Problem 3 (30%) Problem using the Δ - Y transformation

For the circuit in Fig. 4.34, we have $V_{G1} = 100$ V and

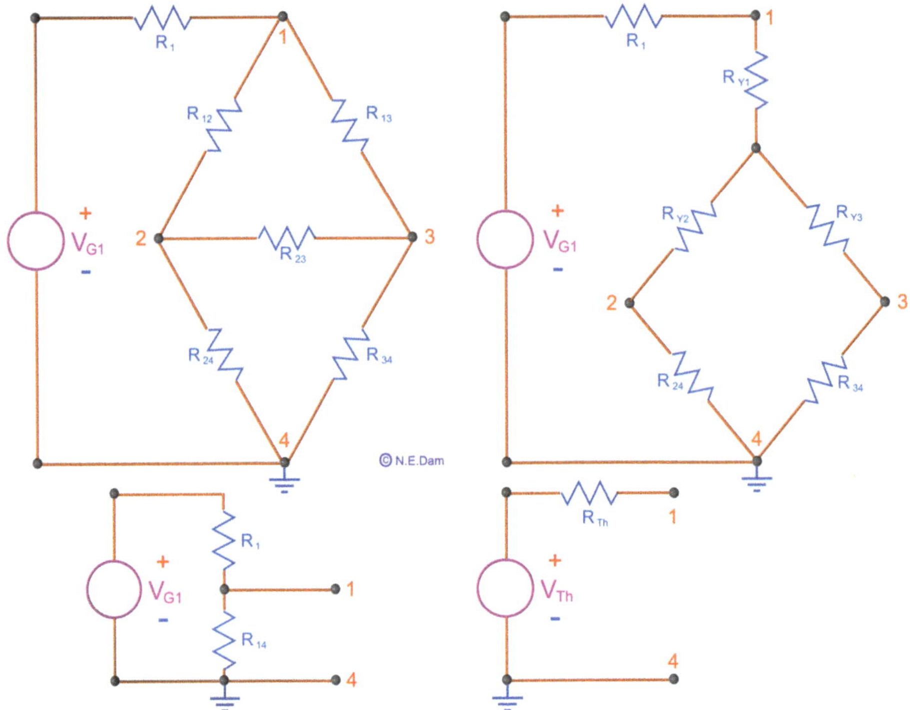

Fig. 4.34 Problem to be solved using a Δ - Y transformation

$$R_1 = 74\ \Omega,\ R_{12} = 20\ \Omega,\ R_{13} = 30\ \Omega,\ R_{23} = 50\ \Omega,\ R_{24} = 30\ \Omega,\ R_{34} = 25\ \Omega$$

A. You should transform the triangle 1-2-3 into a Y-connected circuit and write the general expressions (without figures) for $R_{Y1},\ R_{Y2},\ R_{Y3}$ connected to nodes 1, 2, 3.
B. Use component values and find the equivalent resistance R_{14} between nodes 1 and 4.
C. What type of circuit is $V_{G1},\ R_1$ and R_{14} ?
D. Find the potential V_1 at node 1 if $V_4 = 0$ V.
E. Now, we consider nodes 1 and 4 to be the output terminals of the circuit in question 3. Apply the Thévenin rules and find[2] the equivalent Thévenin circuit.
F. Find the equivalent Norton circuit.[3]

[2] Perhaps $\frac{100}{74} = 1,3514$ and $\frac{26}{1,3514} = 19,24$ and $74 \cdot 26 = 1924$ may be helpful.
[3] Perhaps $\frac{1}{19,24} = 0,052$ may be useful.

Fig. 4.35 Problem to be solved using the Superposition method

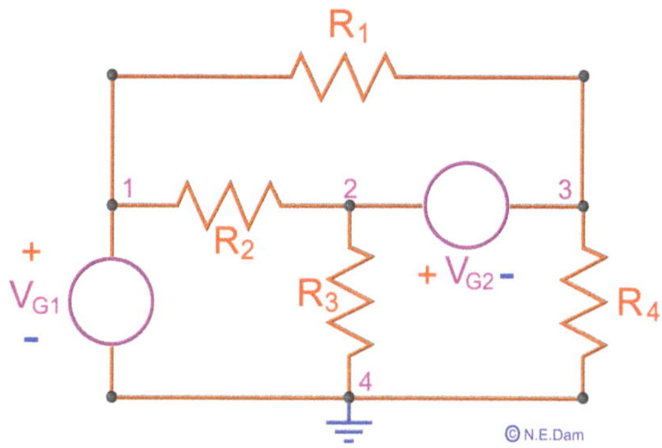

4.15.5 Test 4.5 The Superposition Method

Problem 1 (100%) Solve this problem using the Superposition method

For the circuit in Fig. 4.35, we have:
$R_1 = 2\ \Omega,\ R_2 = 4\ \Omega,\ R_3 = 6\ \Omega,\ R_4 = 8\ \Omega,\ V_{G1} = V_{G2} = 20\ \text{V}$

A. For the circuit, you should write the equations to find V_2 using the superposition method and explain the method with headlines and sketches. You should not write component values on the sketches; only write $R_1,\ R_2,\ R_1 \| R_2,\ V_{G1}$, etc.
B. Use component values and find V_1 and V_2 using[4] the result of Question 1.

[4] You may use $\frac{4}{3} = 1,33,\ \frac{48}{14} = \frac{24}{7} = 3,43$ and $\frac{3,43}{4,76}20 = 14,41$..

Glossary

Glossary of key terms in Danish

antiderivative	stamfunktion.
base a	grundtal a.
base of the natural logarithm	den naturlige logaritmes grundtal.
definite integral	et bestemt integral.
Delta-connected circuit	trekant koblet kredsløb.
improper fraction	uægte brøk.
indefinite integral	ubestemt integral.
logarithm	$\log_b(x)$ er logaritmen med grundtal b til x.
minor	underdeterminant.
power function	potens funktion.
proper fraction	ægte brøk.
sign factor	fortegnsfaktor.
sub-determinant	underdeterminant.
Y-connected circuit	stjerne koblet kredsløb.

N. Dam, *An Introduction to Electrical Circuit Theory*, Synthesis Lectures on Engineering, Science, and Technology, https://doi.org/10.1007/978-3-032-34346-8

Index

N. Dam, *An Introduction to Electrical Circuit Theory*, Synthesis Lectures on Engineering, Science, and Technology, https://doi.org/10.1007/978-3-032-34346-8

Kolibri 360 GmbH
Ferdinand-Jühlke-Straße 7
99095 Erfurt, Deutschland
produktsicherheit@kolibri360.de